Helpline in Biochemistry

(Questions and Answers)

(Second Edition)

Harbans Lal
Senior Professor of Biochemistry
Pt. B.D. Sharma University of Health Sciences
Rohtak (Haryana)

PEEPEE
PUBLISHERS AND DISTRIBUTORS (P) LTD.

Helpline in Biochemistry (Questions and Answers)

Published by
Pawaninder P. Vij

Peepee Publishers and Distributors (P) Ltd.
Head Office: 160, Shakti Vihar, Pitam Pura
Delhi-110 034 (India)

***Correspondence Address:* 7/31, First Floor, Ansari Road
Daryaganj, Post Box-7243, New Delhi 110002 (India)**

Ph: 65195868, 23246245, 9811156083

e-mail: peepee160@yahoo.co.in

e-mail: peepee160@rediffmail.com

e-mail: peepee160@gmail.com

www.peepeepub.com

First Edition: **2005**
Second Edition: **2008 (Oct.)**

ISBN: 81-8445-050-8

Printed at
Lordson, C-5/19, Rana Pratap Bagh, Delhi-110 007

Preface to the Second Edition

I am highly thankful to all the readers of the first edition of this book **'Helpline in Biochemistry (Questions and Answers)'** for giving me this opportunity to present its second edition before you. Interaction with the faculty and the students of the various Medical, Dental and Pharmacy colleges has provided useful suggestions which have helped me for its improvement and I am highly thankful to each one of them for the same. Accordingly, some of the chapters have been modified. New chapters on integration of metabolism, intracellular matrix, free radicals, and hemoglobin metabolism have been added.

Efforts of Mr. Pawaninder P. Vij, Director, Peepee Publishers and Distributors (P) Ltd., New Delhi, deserve special mention and I am highly thankful to him and his team for bringing out this edition.

Harbans Lal

Preface to the First Edition

I am happy to present this book **"Helpline in Biochemistry (Questions and Answers)"**, where the text has been written with an effort to attempt short notes as well as other types of questions which a student has to write in various examinations. The book has been written in a very lucid manner highlighting recent aspects of the subject. It is expected that it will be beneficial for the students of MBBS, BDS, Pharmacy, Physiotherapy, Nursing and all others who are to study Biochemistry in relation to Medicine.

I am highly thankful to all my colleagues and friends without whose efforts and help it was not possible to complete this book. I am also thankful to Mr Pawaninder P Vij, Director, Peepee Publishers and Distributors (P) Ltd., New Delhi, for bringing out this book.

Though every effort has been made to minimise mistakes, I welcome comments and suggestions from my fellow teachers as well as all those who use this book as part of their studies.

Harbans Lal

Contents

1

Cell Organelles and Biological Membranes

Q. Differentiate between a Prokaryotic and a Eukaryotic cell.

Ans.

Prokaryotic cell	*Eukaryotic cell*
1. Prokaryotic cells are unicellular, *e.g.* bacteria	1. Eukaryotic cells may be unicellular as well as multicellular, *e.g.* hepatocyte, erythrocyte, etc.
2. These are small in size, varying from 1–10 nm in diameter	2. These are approximately 10,000 times large and more complex in structure
3. They have only a single membrane which is usually surrounded by a cell wall	3. These have cell membrane and several other membrane containing intracellular components, *e.g.* nucleus, mitochondria, Golgi apparatus, etc.
4. These are similar in structure within one species	4. They vary from one tissue to another with respect to their structure as well as function, *e.g.* liver parenchymal cells, nerve cells, RBC, etc.
5. There is a single chromosome and a molecule of double helical DNA which is known as nuclear *zone*.	5. It contains several chromosomes which get separated into daughter chromosomes during mitosis.

Q. What are ribosomes ?

Ans. Ribosomes are ribonucleoprotein particles containing RNA and protein. These are present on the rough endoplasmic reticulum as well as free particles in the cytoplasm. These are the sites of protein synthesis.

Q. Differentiate between rough endoplasmic reticulum and smooth endoplasmic reticulum.

Ans. Rough endoplasmic reticulum is also called as granular type of endoplasmic reticulum since it has small granules attached to it. These granules are called ribosomes.

Smooth endoplasmic reticulum is the agranular type of endoplasmic reticulum as it consists of membranous structure only and does not contain ribosomes on it.

Q. What is a nucleus ?

Ans. A nucleus is present in an eukaryotic cell. It is the largest component of the cell and contains DNA which is organised into chromosomes. Nucleus is filled with nucleoplasm which has several nucleoli and thread like structure called chromatin.

Q. Describe important functions of nucleus ?

Ans. Important functions of the nucleus include DNA replication, *i.e.* the control of cell division and protein synthesis (by controlling the synthesis of RNA).

Q. What are mitochondria ? Explain their important functions.

Ans. Mitochondria are called as power house of the cell since their major function is to produce energy, in the form of ATP. They have their own DNA which can transcribe its own RNA that translate into proteins.

Mitochondria are rich in enzymes for citric acid cycle and β-oxidation.

Q. What are lysosomes ?

Ans. Lysosomes are small sac-like organelles, rich in all types of hydrolases such as glycosidases, lipases, proteases and nucleases which are required for the digestion of carbohydrates, lipids, proteins and nucleic acids, respectively.

Q. What are suicide bags ?

Ans. Lysosomes are called suicide bags, as these particles are rich in various hydrolytic enzymes. Any disruption to their membrane results in the release of these enzymes within the cell which in turn leads to the digestion of their own components (autolysis of the cell).

Q. Define residual bodies.

Ans. Residual bodies are the vesicles containing large number of ingested material such as foreign particles, dead bacteria or damaged cellular components.

Residual bodies contain high concentration of lipids and are also referred to as lipofuschin. As they accumulate in the cells of the older individuals, these are also referred to as age-pigments.

Q. What is cell membrane ?

Ans. Cell membrane is an outermost covering of the cell which acts as a selective barrier for the transport of substances and separates a cell from the other. It is made up of lipids and proteins.

Q. Describe the composition of a biological membrane ?

Ans. Biological membranes are composed of lipids, proteins and carbohydrates. Different types of membranes have different composition.

A eukaryotic cell may have cell membrane as well as membranes of the other cell organelles such as nuclei, mitochondria, etc. These are called intracellular membranes.

Q. What is a micelle ?

Ans. A micelle is a spheroidal aggregate where the amphiphilic molecules, *e.g.* soap, detergent, etc. are arranged in a way that their hydrophilic groups interact with the aqueous solvent while the hydrophobic groups are in the center.

Q. What are lipid bilayers ?

Ans. A lipid bilayer exists as a sheet in which the hydrophobic regions of phospholipids are protected from the aqueous environment while hydrophilic regions emerge in water. These are extremely stable structures found in a biological membrane.

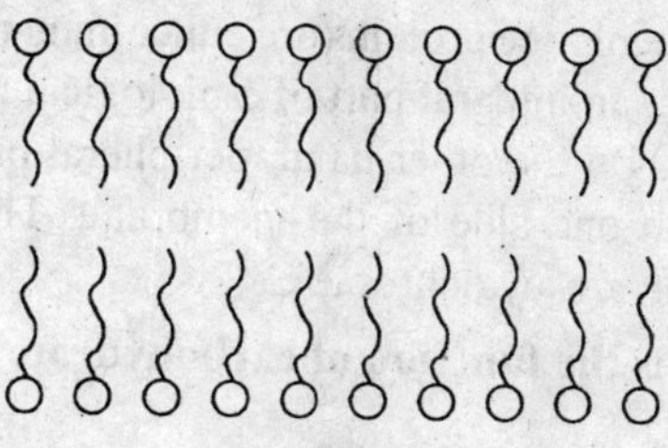

Lipid bilayer.

Q. What are liposomes ?

Ans. A lipid bilayer when encloses on itself and forms a spherical vesicle which separates the external environment from the internal compartment, is called as a liposome. Liposomes have a diameter of several hundred mμ and have been used as the carriers for drug delivery to the target organ.

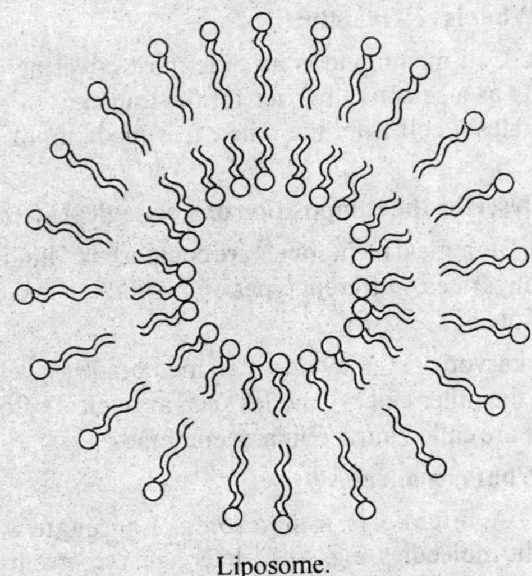

Liposome.

Q. Describe different types of proteins present in a biological membrane.

Ans. Protein concentration in a biological membrane may vary from 20 to 70%, depending upon its function in the cell.

Membrane proteins may be important as structural component, transport systems, enzymes or as hormone receptors. These proteins may be present as integral proteins or peripheral proteins.

Integral membrane proteins are either immersed partially or totally and form an integral part of a biological membrane, *e.g.* glycophorin A. On the other hand, peripheral proteins are embedded only on one side of the membrane. They are usually soluble in water, *e.g.* cytochrome C.

Q. What is the function of carbohydrate in a biological membrane ?

Ans. The carbohydrate content of a membrane may vary from 3 to 10%. These are present as oligosaccharides and are covalently attached either to a protein (as glycoprotein) or to a lipid (as glycolipid). The oligosaccharide chains are located normally on the outer surface of the membrane for signal transduction.

Q. Describe the fluid-mosaic model of a cell membrane ?

Ans. Fluid-mosaic model of a biological membrane explains that a membrane consists of mosaic of proteins which are free to drift in the plane of lipids, *i.e.* biological membranes have such a structure where some proteins span the lipid bilayer whereas others are only partially immersed.

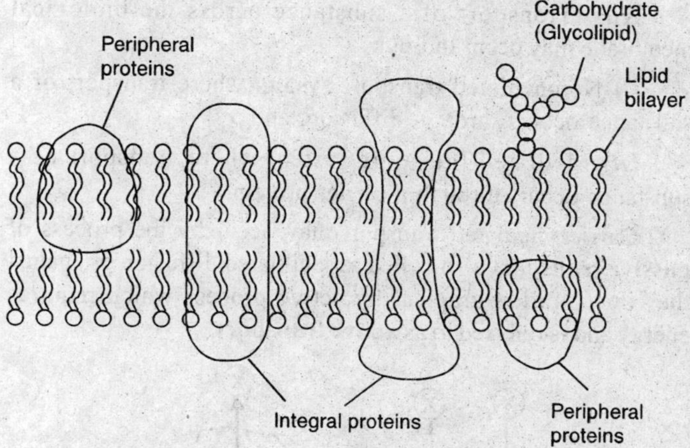

Q. What is fluidity of a membrane ?

Ans. The fluidity of a membrane is due to the presence of unsaturated fatty acids. Due to the presence of the double bond, the fattyacyl chain bends and forms kink, as a result of which the membrane can not be tightly packed and becomes more fluid in nature.

Fluidity of the membrane affects its functions such as control of activity of membrane bound enzymes, cell growth, etc.

Q. Describe functions of biological membrane proteins.

Ans. Membrane proteins have various functions—

(*i*) Some of these are structural proteins and influence cell shape or movement.

(*ii*) Several membrane proteins regulate the transport of certain molecules such as glucose and amino acids.

(*iii*) Many of the membrane proteins are enzymes and are located either on the inner surface of the membrane or within the membrane.

(*iv*) Several membrane proteins act as recognition sites for various types of molecules such as hormone receptors, LDL receptors or immune receptors, etc.

(*v*) Various membrane proteins help in transmission of signals such as G-proteins, adenylate cyclase, etc.

Q. Describe various transport systems.

Ans. Transport of a substance across the biological membrane may occur through —

(*i*) Nonmediated-transport system where transport of a substance occurs through diffusion, and

(*ii*) Mediated-transport system where transport of a substance occurs through a carrier protein.

Carrier-mediated transport may occur by the process of **passive transport** also called as facilitated diffusion or against the concentration gradient by active process which requires energy and is referred to as **active transport**.

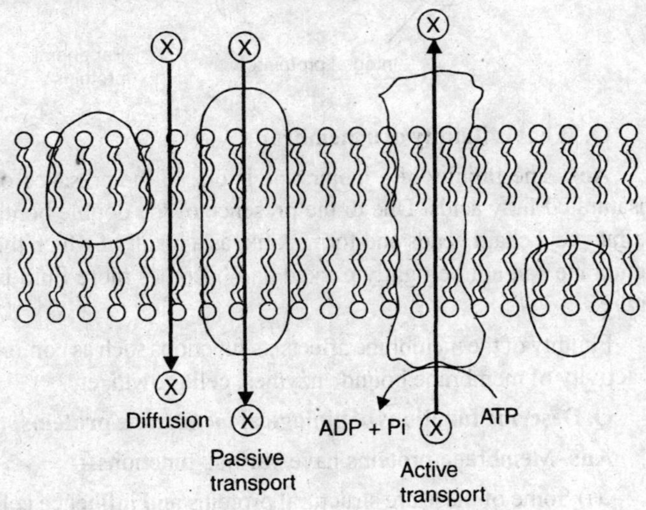

Transport systems.

Q. Describe uniport, symport and antiport transport of molecules.

Ans. (*i*) **Uniport transport** – movement of one types of molecules occur unidirectionaly, *e.g.* transport of glucose in erythrocytes.

(*ii*) **Cotransport** – movement of one type of solute depends upon the simultaneous movement of the other.

It may further be divided into—

Symport where two types of molecules move in the same direction, *e.g.* Na^+ – glucose transport, and

Antiport where two types of molecules move in opposite directions *e.g.* Na^+ – K^+ transport.

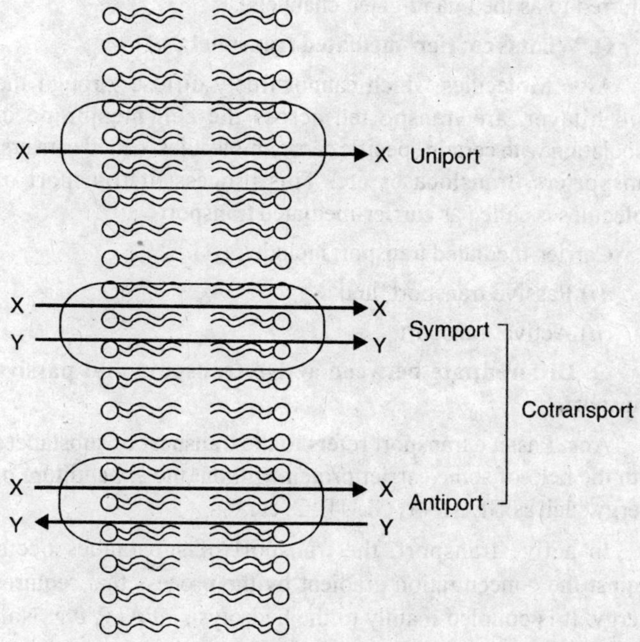

Uniport, Symport and Antiport Transport of Molecules.

Q. What is nonmediated transport ?

Ans. Nonmediated transport is the transport of substances such as O_2, CO_2, NO, etc., across the cell membrane, by diffusing down the electrochemical gradient. This process does not require metabolic energy and is also called as diffusion.

Diffusion may occur through transmembrane routes such as via channels and pores.

Q. What are membrane channels ?

Ans. These are transmembrane channels composed of proteins which allow rapid movement of ions and molecules from one side of the membrane to the other. These channels have passage (gate) and the flow of ions/molecules in turn regulates opening or closing of the channels. This involves a conformational change in a membrane protein which may either be due to a change in voltage membrane potential (called as the voltage-gated channel) or in response to the binding of a ligand such as cyclic nucleotides (referred to as the ligand-gated channels).

Q. What is carrier–mediated transport ?

Ans. Molecules which cannot freely diffuse through the lipid bilayer, are transported across the cell membrane in association with certain specific carrier molecules, called carriers, transporters, translocases etc. This process of transport of molecules is called as carrier-mediated transport.

Carrier-mediated transport includes :

(*i*) Passive transport, and

(*ii*) Active transport.

Q. Differentiate between active transport and passive transport.

Ans. Passive transport refers to the transport of substances with the help of some carrier protein without the expenditure of energy such as GLUT– 1, GLUT– 2, etc.

In active transport, the transport of substances occur against the concentration gradient by the process that requires energy. It is coupled mainly to the hydrolysis of ATP, *e.g.* Na^+, K^+–ATPase.

Q. Describe some transport disorders.

Ans. A pathological condition due to a defect in membrane transport system, for a specific cellular component, is called as a transport disorder. Some of the transport disorders are —

(*i*) **Hartnup's disease :** In this condition there is impaired transport of neutral amino acids including tryptophan either in the epithelial cells of the intestine or the tubular cells.

(*ii*) **Cystinuria** is due to impaired renal reabsorption of cystine and other basic amino acids.

Q. What is group translocation ?

Ans. This is the process of transport of amino acids across the plasma membrane of some tissues with the help of glutathione.

It involves chemical modification of the transported substance. The substrate, such as glutathione, and the amino acid are released into the cell with certain modifications.

Q. What is endocytosis ?

Ans. Endocytosis is a mechanism for the transport of large molecules such as proteins, intracellularly. The molecule to be transported is enclosed within the plasma membrane which in turn forms a vesicle, moves into the interior of cell and delivers its contents, intracellularly.

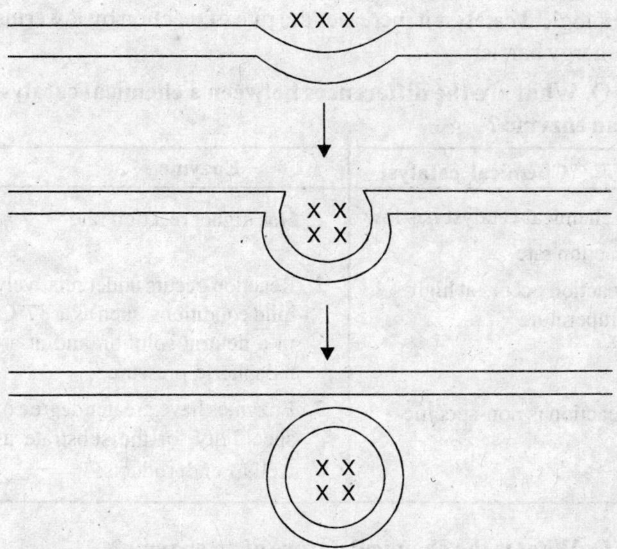

Q. What is exocytosis ?

Ans. Exocytosis is the process of the release of large molecular weight substances from inside to the outside of the cell. Several proteins are transported extracellularly by this process.

2

Enzymes

Q. What are enzymes ?

Ans. Enzymes are biological catalysts, mostly proteinaceous in nature. These are nondialysable, colloidal organic macromolecules, produced in the living system and catalyze biochemical reactions without being utilized in the process. Since, an enzyme is a biological catalyst it increases the rate of reaction by lowering free energy barrier.

Q. What are the differences between a chemical catalyst and an enzyme ?

Ans. **Chemical catalyst**	**Enzyme**
1. A chemical catalyst has low reaction rate	1. Has higher reaction rate
2. Reaction occurs at high temperature	2. Reaction occurs under relatively mild conditions, such as at 37°C, in a neutral solution and at atmospheric pressure
3. Reaction is non-specific	3. Enzymes have greater degree of specificity for the substrate as well as endproducts.

Q. What is the chemical nature of an enzyme?

Ans. Mostly, enzymes are thermolabile proteins with large molecular weight though some of the enzymes may be RNA in nature (ribozymes).

Q. What is a Ribozyme ?

Ans. A ribozyme is a RNA molecule with high substrate specific catalytic activity, *e.g.* RNAase–P from E.coli.

These are important in the conversion of pre-mRNAs to mature mRNAs.

Q. What are oligomeric enzymes ?

Ans. Enzymes which have multiple subunits are called as

oligomeric enzymes, *e.g.* lactate dehydrogenase. It is a tetramer and has four subunits.

Q. What are multienzyme complexes ?

Ans. An enzyme with more than one activity is called as multienzyme complex. Such an enzyme complex catalyzes the conversion of a substrate forming different products, sequentially, *e.g.* fatty acid synthase. It is a multienzyme complex having seven different enzyme activities.

Q. What is a zymogen ?

Ans. Zymogens are enzymes which are secreted as large molecules called proenzymes. They undergo proteolysis and are converted to their biologically active form, *e.g.* proteolytic enzymes like pepsin, trypsin, etc.

Pepsin is synthesized as pepsinogen (a zymogen). In the presence of HCl (from the gastric juice), pepsinogen is converted to pepsin (active enzyme).

Once some pepsin has been synthesized, it can further activate pepsinogen. This process is called autocatalysis.

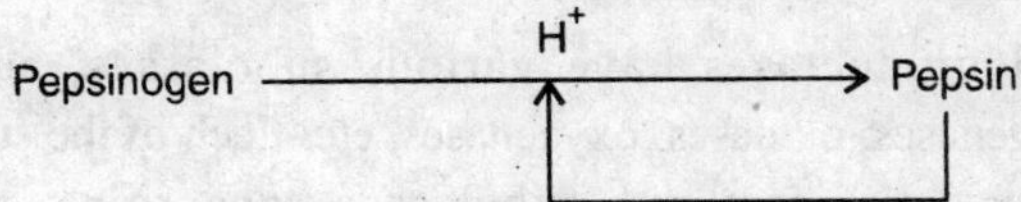

Q. Classify enzymes.

Ans. Enzymes are classified into 6 major classes.

1. Oxidoreductases, *e.g.* alcohol dehydrogenase, succinate dehydrogenase
2. Transferases, *e.g.* hexokinase, glutamate pyruvate transaminase
3. Hydrolases, *e.g.* glucose 6-phosphatase, pepsin
4. Lyases, *e.g.* histidase, aldolase
5. Isomerases, *e.g.* phosphohexose isomerase, phosphoglucomutase,
6. Ligases, *e.g.* glutamine synthetase, glutathione synthetase.

This classification is based on the type of reaction, an enzyme catalyses.

According to the enzyme commission of International Union of Biochemistry and Molecular Biology, each enzyme is assigned a four digit code, called as enzyme commission number (EC number).

The first digit of the EC number denotes the main class,

the second digit characterizes the subclass,

the third digit indicates sub-subclass while

the fourth digit denotes serial number of a particular enzyme in each sub-subclass, *e.g.* 1.1.1.1 indicates alcohol dehydrogenase. It is the first enzyme of the sub-subclass I which uses NAD as hydrogen acceptor. It belongs to subclass I, *i.e.* dehydrogenases and belongs to class I, *i.e.* oxidoreductases.

Q. What are oxidoreductases?

Ans. Oxidoreductase is the class of enzymes which catalyses oxidation//reduction of their substrates by the addition of oxygen or removal of hydrogen/electrons, *e.g.* succinate dehydrogenase, phenylalanine hydroxylase, etc.

Oxidoreductases have various subclasses, such as dehydrogenases, oxidases, oxygenases, etc. Each of the subclass may have different sub-subclasses, according to the co-enzyme required, *e.g.* enzymes requiring NAD^+, $NADP^+$, FAD^+, etc.

Q. What are transferases ?

Ans. Transferases are a group of enzymes that catalyse transfer of a functional group from one substrate to another. Transferases may be subclassified as aminotransferases (gluta-mate oxaloacetate transaminase), phosphotransferases (hexokinase), etc.

Q. What are hydrolases ?

Ans. Hydrolases is a group of enzymes by which a substrate is spiltted into two products, with the addition of a molecule of water. They are subclassified with respect to the nature of the group or bond hydrolysed, *e.g.* peptidases (pepsin), glycosidases (amylase), lipases, etc.

Q. What are lyases ?

Ans. Lyases catalyse the removal of a group (water, ammonia, CO_2, etc) from a substrate without the addition of a molecule of water, *e.g.* fumarase, etc.

Q. What are isomerases ?

Ans. Isomerases catalyse the interconversion of one isomeric form of a compound to the other, e.g. phosphohexose-isomerase, phosphotrioseisomerase, etc.

Q. What are ligases ?

Ans. Ligases are enzymes which catalyse the synthesis of a product by joining two substrates. Most of these enzymes catalyse a reaction at the expense of energy from ATP, *e.g.* glutamine synthetase. Such enzymes are called **synthetases.**

Ligases may also result in the ligation of two substrates without the expenditure of ATP, *e.g.* glycogen synthase. Such enzymes are called **synthases.**

Q. What is a cofactor ?

Ans. Some of the enzymes require the presence of certain small molecules, *e.g.* a metal ion or an organic molecule such as NAD, for their activity. These molecules are called **cofactors**. Organic molecules such as NAD, FAD, etc, may also be referred as coenzymes.

Q. What are coenzymes ?

Ans. Coenzymes are organic molecules required with certain enzymes to catalyse a particular reaction. They may act either as cosubstrate or a prosthetic group.

The protein part of the enzymes called **apoenzyme**. While the prosthetic group is called **coenzyme**. The protein-prosthetic group complex (apoenzyme + coenzyme) is called **holoenzyme**.

Apoenzyme	+	Coenzyme	=	Holoenzyme
(Protein)				(Active enzyme)

(Biologically inactive)

Coenzymes are mainly derivatives of the B complex group of vitamins, *e,g,* TPP, a derivative of thiamin.

Q. What are abzymes ?

Ans. Abzymes are artificially synthesized catalytic antibodies which are produced against an enzyme-substrate complex, in the transition state of the reaction.

Q. Describe various factors affecting enzyme activity?

Ans. An enzyme shows maximum activity only under optimal conditions. Activity of an enzyme can be affected by several factors :—

1. Temperature
2. pH
3. Substrate concentration
4. Concentration of the activator or inhibitor

Q. What is the effect of temperature on enzyme activity?

Ans. Rate of enzyme catalyzed reaction increases with the increase in temperature. It is maximum at optimum temperature and falls thereafter, exhibiting a bell-shaped curve.

An increase in the rate of a reaction with the increase in temperature is due to the available energy of activation of the system. After the enzyme activity has reached its maximum at the optimum temperature (which is approximately 37° C for most of the enzymes), activity is reduced due to denaturation at high temperature.

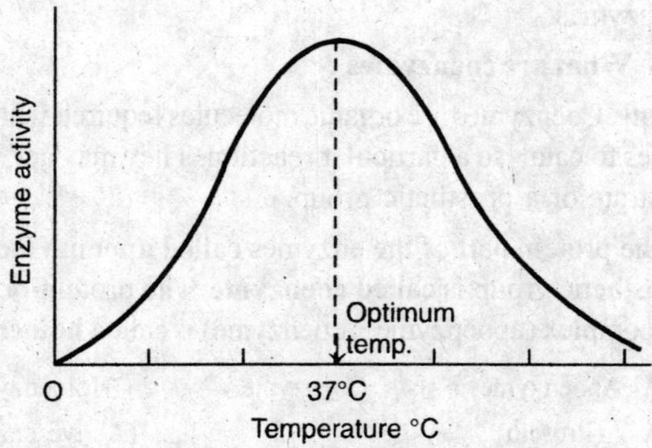

Q. What is the effect of pH on enzyme activity ?

Ans. An enzyme-substrate complex is formed due to ionization of certain groups on the enzyme as well as the substrate. A change in pH affects ionization of the enzyme (protein) thereby decreasing the number of active sites.

Extremely high as well as low pH denature the enzyme and each enzyme has its optimum pH (though most of the enzymes best function at physiological pH). Thus an increase in pH increases enzyme activity which is maximum at optimum pH and is again reduced thereafter with the increase in pH, hence showing a bell-shaped curve.

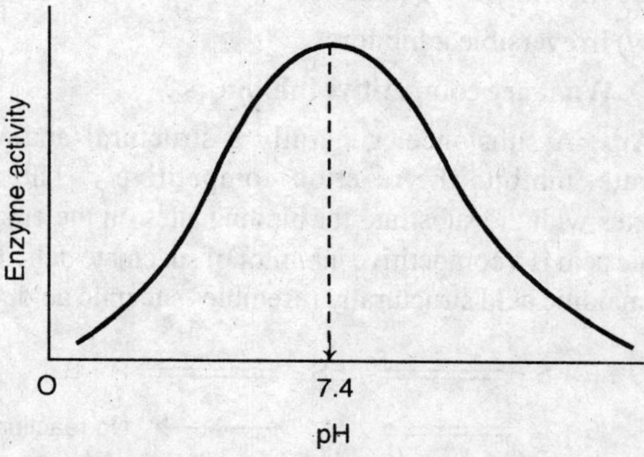

Q. What is the effect of substrate concentration on enzyme activity ?

Ans. For an enzyme catalyzed reaction (with a single substrate and a single product), rate of the reaction is directly proportional to its substrate concentration upto a point where enzyme gets saturated.

Reaction velocity reaches a maximum value and remains unchanged thereafter.

Q. What is Michaelis constant [Km] ?

Ans. Km is defined as the substrate concentration at which reaction velocity (V) is half of the maximum velocity (Vmax/2).

Q. What is the effect of enzyme concentration on reaction velocity ?

Ans. The rate of an enzyme catalyzed reaction is directly proportional to the concentration of the enzyme.

Q. What are enzyme inhibitors ?

Ans. A substance which inhibits enzyme activity and in turn reduces the velocity of an enzyme catalyzed reaction is called inhibitor, *e.g.* malonic acid is an inhibitor of succinate dehydrogenase.

Q. Name different types of enzyme inhibitors ?

Ans. Enzyme inhibitors are divided into different groups; as follows –

(*a*) Competitive inhibitors

(*b*) Non-competitive inhibitors

(*c*) Uncompetitive inhibitors

(*d*) Irreversible inhibitors.

Q. What are competitive inhibitors ?

Ans. A substance, generally a structural analog of the substrate, inhibits the reaction competitively. The inhibitor competes with the substrate for binding sites on the enzyme, *e.g.* malonic acid is a competitive inhibitor of succinate dehydrogenase since malonic acid structurally resembles succinic acid.

$$E + S \rightleftharpoons ES \rightleftharpoons P + E$$

$$E + I \rightleftharpoons EI \not\longrightarrow \text{No reaction}$$

It is a reversible process as degree of inhibition can be reduced by the increasing concentration of the substrate.

The presence of a competitive inhibitor increases Km of the substrate without any change in Vmax.

Certain drugs such as sulfonamide is a competitive inhibitor of p-aminobenzoic acid. Similarly, Methotrexate (a structural analog of folate) is a competitive inhibitor of dihydrofolate reductase.

Q. What are non-competitive inhibitors ?

Ans. Such an inhibitor has no structural similarity to the substrate and can form a complex either with the enzyme or with the enzyme-substrate complex :

$$E + S \rightleftharpoons ES \longrightarrow P + E$$

$$E + I \rightleftharpoons EI \not\longrightarrow \text{No product}$$

$$E + S \longrightarrow ES \xrightarrow{I} ESI \not\longrightarrow \text{No product}$$

Such type of inhibitors decrease Vmax but do not change the Km, *e.g.* 5-fluorouracil. It is a noncompetitive inhibitor of thymidylate synthetase.

Q. What are irreversible inhibitors ?

Ans. Irreversible inhibitors includes heavy metals, poisons and several oxidizing agents. Such compounds do not have structural similarity with the substrate. These type of inhibitors result in chemical modification of the enzyme, irreversibly, *e.g.* cyanide, lead, etc. These inhibitors are also called **inactivators**. Irreversible inhibitors reduce Vmax without changing the Km.

Q. What are allosteric enzymes ?

Ans. The oligomeric enzymes (having two or more subunits) have an additional site other than the substrate-binding site. This site is referred to as catalytic site. Such enzymes are called **allosteric enzymes** or **regulatory enzymes**, since they are allosterically regulated. The substances which regulate such enzymes are called allosteric modulators. An allosteric modulator may activate an enzyme (allosteric activator) or inhibit the enzyme (allosteric inhibitor). HMG CoA reductase is a regulatory enzyme of cholesterol biosynthesis.

When reaction velocity is plotted against substrate concentration, allosteric enzymes show sigmoidal curve. Activity of such an enzyme is regulated by covalent modification.

Q. What is feed-back inhibition ?

Ans. Feed-back inhibition refers to the inhibition of an enzyme by the end product of the pathway.

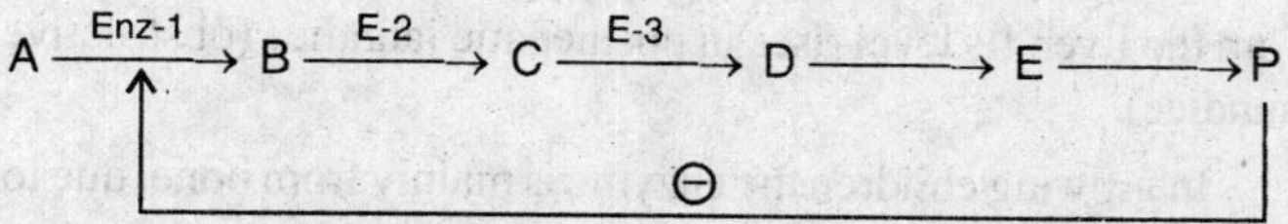

Accumulation of the end product slows down the whole reaction sequence and as the product is utilized, the reaction continues.

Q. What are isoenzymes ?

Ans. Isoenzymes, also called **isozymes**, are physically distinct form of an enzyme with the same specificity. They however, differ from each other with respect to their source, electrophoretic mobility and imunological properties, *e.g.* lactate dehydrogenase, creatine phosphokinase, etc.

Isozymes have wide clinical importance in diagnosis.

Q. Name different isozymes of LDH ?

Ans. LDH is a tetramer having four different subunits of the two subtypes referred to as H type and M type. Four different subunits give rise to five different isoenzymes. LDH_1 has four H subunits (H_4) and is found in the myocardium while LDH_5 has four M subunits (M_4) and is predominantly present in the liver and the skeletal muscle. Accordingly, LDH_1 and LDH_2 are of diagnostic significance in cardiac diseases whereas LDH_4 and LDH_5 have diagnostic significance in liver diseases.

Q. What are isozymes of CPK ?

Ans. CPK, also called as creatine kinase (CK), has two subunits, *i.e.* M type and B type. It thus exists in three different forms, *i.e.* as CPK-1 (BB) found in the brain, CPK-2 (MB) in the myocardium and CPK-3 (MM) in the skeletal muscle.

CPK-2 is normally present in very small amount in the serum. Its level rises with in four hours of myocardial infarction and reaches to a maximum with in 24 hours.

Q. Describe various isozymes of alkaline phosphatase.

Ans. Alkaline phosphatase exists in three different forms. These are named according to the source of origin *i.e.* the liver, bone and placenta.

A large portion of alkaline phosphate in serum is derived from the liver. Its level rises in posthepatic jaundice (obstructive jaundice).

In growing children the enzyme is mainly from bone, due to high osteoblastic activity.

The third form of the enzyme is called heat stable alkaline phosphatase. It is present in the serum in females during the third trimester of pregnancy and is of **placental** origin.

Q. Describe diagnostic significance of enzymes.

Ans. Serum levels of several enzymes are raised in various clinical conditions and are helpful in the diagnosis/prognosis of a disease, *e.g.* aspartate aminotransferase (SGOT), alanine aminotransferase (SGPT), amylase, LDH, CPK, etc.

The table shows the diagnostic significance of various enzymes in different diseases –

ENZYME	CLINICAL CONDITIONS IN WHICH ALTERED
1. Acid phosphatase	Carcinoma of the prostate
2. Alkaline phosphatase	Obstructive jaundice, bone diseases
3. Amylase	Acute pancreatitis
4. CPK	Myocardial infarction, muscular dystrophy
5. LDH	Myocardial infarction, acute hepatitis, muscle disorders.
6. GOT	Cardiac disorders
7. GPT	Viral hepatitis
8. γ-GT	Alcoholic liver disease.

3

Biological Oxidation

Q. What do you understand by the term biological oxidation ?

Ans. Biological oxidation refers to the phenomenon which extracts energy from food (nutrients) and makes it available to the organism. Since this process requires oxygen, biological oxidation thus refers to the process of oxidation of nutrients in a biological system.

Q. What is a high energy compound ?

Ans. A compound when undergoes an oxidative reaction, releases some amount of energy in the reaction system which can either be captured for an anabolic reaction or used for the phosphorylation of ADP, *e.g.* creatine phosphate, phosphoenol-pyruvate, ATP, GTP, etc.

When a phosphate bond of ATP is hydrolysed it releases 7.3 Kcal of energy.

$$\text{ATP} \longrightarrow \text{ADP + Pi} \quad (\searrow 7.3\ \text{Kcal})$$

Q. What are reducing equivalents ?

Ans. A reducing equivalent may be defined as a proton or electron $[H^+ + e^-]$. Electrons are generated when a substrate is oxidized.

Reducing equivalents are used for the production of energy (generation of ATP) via electron transport chain. Lipids are highly reduced substances compared to carbohydrates or proteins.

Q. Name some enzymes which are used in the process of biological oxidation ?

Ans. The process of biological oxidation involves either the addition of oxygen or removal of hydrogen/electrons with the help of various oxidoreductases, *e.g.* NADH dehydrogenase, succinate dehydrogenase, cytochrome oxidase, etc.

Q. Describe various components of the electron transport chain ?

Ans. Various components of the electron transport chain are

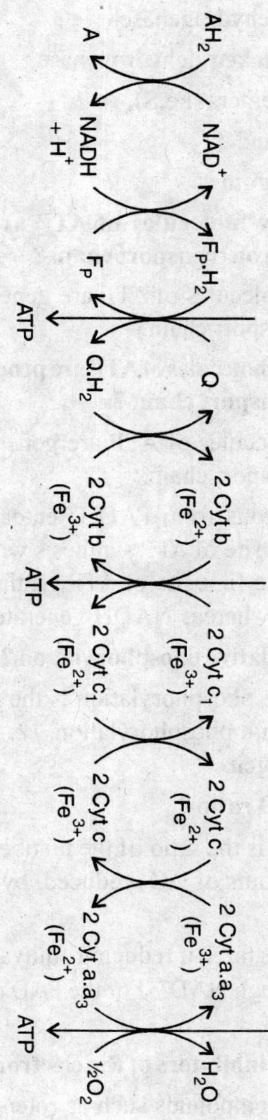

Q. What is electron transport chain ?

Ans. Electron transport chain is the systematic arrangement of various electron carriers from NADH/$FADH_2$ to oxygen.

Various enzymes and proteins which participate in electron transport chain include –

NAD-linked dehydrogenase,

Flavoprotein-linked dehydrogenase,

Iron-sulphur centers (Fe.S),

Coenzyme Q, and

various cytochromes.

Q. How many molecules of ATP are produced when NADH enters electron transport chain ?

Ans. Three molecules of ATP are generated when NADH enters electron transport chain.

Q. How many molecules of ATP are produced when $FADH_2$ enters electron transport chain ?

Ans. Two molecules of ATP are generated when $FADH_2$ enters electron transport chain.

Because electrons from $FADH_2$ enter electron transport chain, after the first site of ATP synthesis whereas from NADH they enter before the first site of ATP synthesis, hence $FADH_2$ generates two ATP whereas NADH generates three ATP.

Q. What is oxidative phosphorylation ?

Ans. Oxidative phosphorylation is the process where oxidation is coupled with phosphorylation, *i.e.* phosphorylation in the presence of oxygen.

Q. What is P/O ratio ?

Ans. P/O ratio is the ratio of the number of ATP generated to the number of atoms of ½O_2 reduced, by the electron transport chain.

The P/O ratio is three if reducing equivalents enter electron transport chain through NAD^+. For the FAD requiring reactions, P/O ratio is two.

Q. Name some inhibitors of the electron transport chain ?

Ans. Various compounds such as rotenone, amytal, barbiturates or cyanide ions inhibit the process of electron transport chain at different points thereby affecting oxygen consumption and generation of ATP.

Rotenone and amytal inhibit the first step of ATP synthesis whereas cyanide or carbon mono-oxide inhibit the terminal step of electron transport chain.

Q. Name the site of inhibition of electron transport chain by cyanide ions. Why it is a most powerful inhibitor of electron transport chain ?

Ans. Cyanide acts at the terminal step of electron transport chain. It binds to Fe^{3+} in the heme of cytochrome oxidase (cytochrome $a.a_3$), preventing the oxygen from reacting with cytochrome oxidase as the final electron acceptor. This in turn results in the arrest of respiration and stoppage of energy production. Cell death occurs rapidly from tissue asphyxia, particularly of the CNS.

Q. Explain the theory of oxidative phosphorylation ?

Ans. In the process of the synthesis of ATP by reducing equivalents and electrons, oxidative phosphorylation is coupled-with electron transport chain. This coupling occurs according to chemiosmotic theory.

Protons, generated during electron transport chain, are transported into the spaces between the inner and the outer membrane. This in turn generates H^+ gradient across the membrane.

In the first step, protons from NADH alongwith two electrons are transferred to flavoprotein of the NADH-dehydrogenase and reduce it. Subsequently, two electrons from $FMNH_2$ are transferred to iron.sulphur centres (Fe.S) which cycles electrons back across the membrane to the interior whereas protons are released into the inner membrane space.

In the next step, two electrons from Fe.S centres are transferred to coenzyme Q , alongwith two H^+ which are derived from the aqueous matrix. Coenzyme Q in turn reduces and translocates H^+ to the exterior surface while the electrons are transferred to cytochrome b. These electrons from cytochrome b are returned to the inner surface where, alongwith two protons from the matrix, they again join coenzyme Q. At this stage, reduced coenzyme Q transfers the third pair of protons ($2H^+$) into the inner membrane space and electrons to cytochrome c_1. From cytochrome c_1 electrons pass on to cytochrome c and are finally accepted by cytochrome a. Within the matrix, these electrons are transferred by cytochrome a_3 to an atom of oxygen.

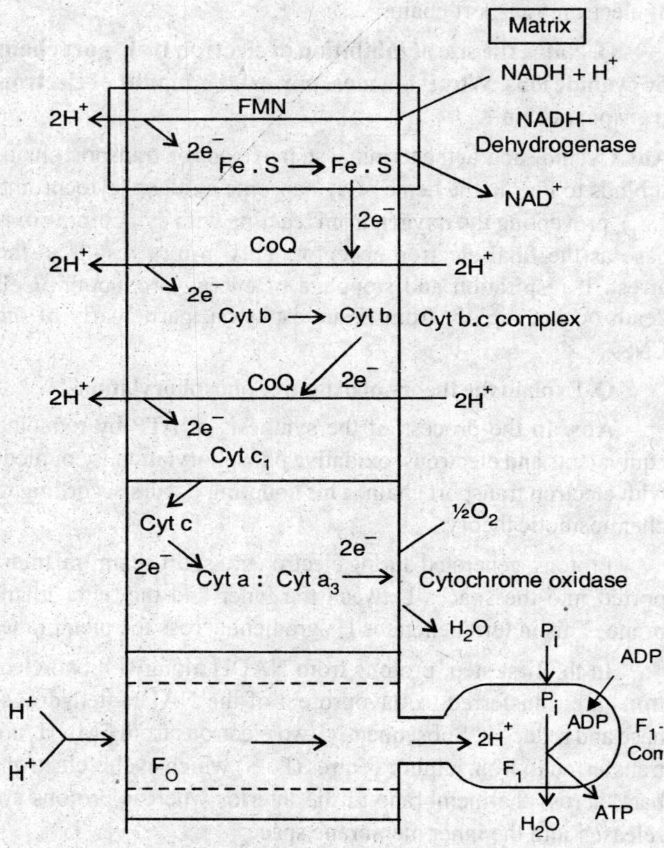

Protons traverse the membrane through F_0 portion of the F_1–F_0 complex, into its end piece, *i.e.* F_1. With the acquisition of electrons by the oxygen, these protons form a molecule of water. The F_1 component also has a binding site for inorganic phosphate near the proton channel. The energy of proton gradient alongwith the membrane potential results in the phosphorylation of ADP to ATP and results in the elimination of water.

Q. Name inhibitors of oxidative phosphorylation.

Ans. Oligomycin is an inhibitor of oxidative phosphorylation sequence of the mitochondrial F_1-F_0 ATPase. This suppresses oxygen consumption and in turn inhibits the process of oxidative phosphorylation and ATP generation.

Q. 15. What is an uncoupler ?

Ans. An uncoupler, *e.g.* 2, 4-dinitrophenol uncouples (inhibits) phosphorylation from electron transport chain. As a result of this, ADP will not be phosphorylated. The process of electron transport will continue uncoupled with the process of oxidative phosphorylation and there will be no generation of ATP.

Q. 16. What is substrate level phosphorylation ?

Ans. It is the process of phosphorylation of ADP to generate ATP without the oxygen. The energy and inorganic phosphate to phosphorylate ADP is directly available from some high energy phosphate donor which is used as a substrate in the reaction, *e.g.* generation of ATP in the muscle from creatine phosphate.

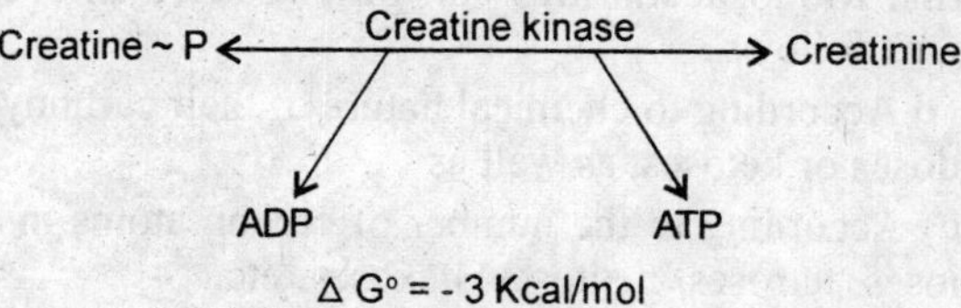

4

Carbohydrates

Q. What are carbohydrates ?

Ans. Carbohydrates are the polyhydroxyaldehydes or ketones containing primarily three elements, *i.e.* carbon, hydrogen and oxygen which are combined as $(CH_2O)_n$, where *n* is 3 or more.

Q. Classify carbohydrates.

Ans. Carbohydrates can be classified into 3 groups :

(*i*) Monosaccharides, *e.g.* glucose,

(*ii*) Disaccharides, *e.g.* lactose, and

(*iii*) Polysaccharides, *e.g.* starch.

Q. What are monosaccharides ?

Ans. Monosaccharides are aldehyde or ketone derivatives of the straight chain polyhydroxyalcohols which contain 3 or more carbon atoms. These are simple sugars which can join together by the glycosidic linkage to form various di and polysaccharides

Q. Classify monosaccharides.

Ans. Monosaccharides can be classified into different groups, as follows –

(*i*) According to chemical nature of their carbonyl group, *e.g.* aldoses or ketoses, as well as –

(*ii*) According to the number of carbon atoms in a chain, *e.g.* trioses, tetroses, pentoses, hexoses, etc.

Q. What is an asymmetric carbon atom ?

Ans. The central carbon atom of an optically active compound whose all the four valencies are occupied by different atoms or groups is referred to as asymmetric carbon atom, *e.g.* :

```
         CHO
          |
H  —  * C  —  OH
          |
        CH2OH
```

D-Glyceraldehyde

(* An asymmetric carbon atom)

Q. What are stereoisomers ?

Ans. An optically active compound rotates the plane of polarized light either to the left or to the right. Such a pair of compounds form nonsuperimposable images of each other. These two molecules are called as stereoisomers or enantiomers of one another, *e.g.* D- and L-glyceraldehyde —

```
      CHO                      CHO
       |                        |
H — *C — OH             HO — *C — H
       |                        |
      CH2OH                    CH2OH
```

D-Glyceraldehyde L-Glyceraldehyde

Q. How many isomers a compound can form ?

Ans. The number of isomers depends upon the number of asymmetric carbon atoms in the molecule. It can be represented by the formula 2^n. For example, glucose has four asymmetric carbons and can have 16 (2^4) stereoisomers.

Q. What is the difference between D/L form and d/l form of a compound ?

Ans. D or L form of a compound represents its spatial relationship with the parent compound of the carbohydrate family, *i.e.* glyceraldehyde.

In Fischer's projection, a molecule of glyceraldehyde with an –OH group at asymmetric center on the right hand side when aldehyde group is at the top, is designated as D-glyceraldehyde.

Its nonsuperimposable mirror image which has this – OH group on the left hand side is referred to as L-glyceraldehyde.

```
     CHO                 CHO
      |                   |
H – C – OH          OH – C – H
      |                   |
     CH2OH               CH2OH
```

D-Glyceraldehyde L-Glyceraldehyde

The *d*/*l* or +/– form of a compound refers to its ability to rotate the angle of the plane of polarized light. An isomer which rotates the plane of polarized light to the right is designated as dextrorotatory (*d* or +) while its isomer which rotates the plane of polarized light to the left is referred to as levorotatory (*l* or –).

An asymmetric molecule with D – configuration can be either dextrorotatory or levorotatory, *i.e.* exhibiting its structural relationship to D-glyceraldehyde but with different optical rotations.

The *d* and *l* (+ and –) isomers are also referred to as optical isomers.

Q. What is a racemic mixture ?

Ans. A mixture containing equal amounts of each of the two isomers of a chiral molecule (asymmetric molecule) is called as racemic mixture (*dl*-mixture).

Q. What are anomers ?

Ans. During the formation of the ring structure of a monossaccharide, the functional carbonyl group at carbon atom one or a keto group (>C = 0) at C_2 gets cyclized and occurs, in two configurations, *i.e.* as the α and the β group. Such a carbon atom is called as an anomeric carbon and the stereoisomers are referred to as anomers, *e.g.* C_1 in glucopyranose or C_2 in fructofuranose. The hexose molecule now has five asymmetric carbon atoms instead of four present in the straight chain structure of the molecule and can exhibit 32 isomers.

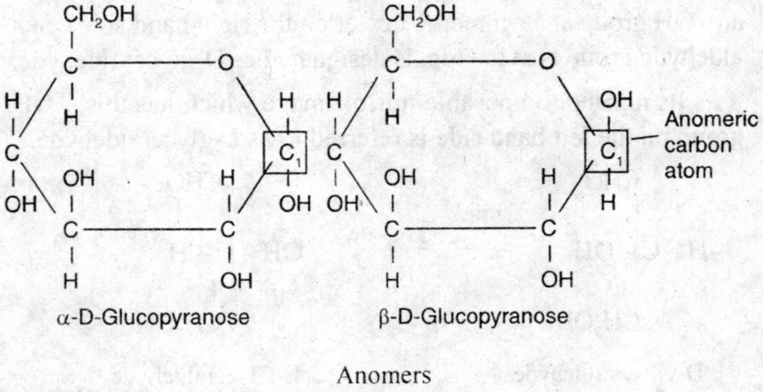

α-D-Glucopyranose β-D-Glucopyranose

Anomeric carbon atom

Anomers

Q. What is mutarotation ?

Ans. Two anomers of glucose are freely interconvertible in an aqueous solution. This interconversion of the two anomers also results in a change in their optical rotation. This phenomenon of change in optical rotation of D (+) glucose in an aqueous solution and reaching a constant value of 52.5°, is called as mutarotation.

α-D-Glucopyranose ⟷ D-Glucose (Aldehyde) ⟷ β-D-Glucopyranose

Q. What are epimers ?

Ans. Sugars that differ by configuration of its H and OH group around only one of the optically active carbon atoms are referred to as epimers, *e.g.* D-mannose differs in configuration around the second carbon atom with respect to D-glucose. Similarly, D-galactose has a different configuration than D-glucose around carbon atom four. Thus D-mannose and D-galactose are the two epimers of D-glucose.

Mannose	Glucose	Galactose
CHO	CHO	CHO
HO – C_2 – H	H – C – OH	H – C – OH
HO – C_3 – H	HO – C_3 – H	HO – C_3 – H
H – C – OH	H – C – OH	HO – C_4 – H
H – C – OH	H – C – OH	H – C – OH
CH_2OH	CH_2OH	CH_2OH

Q. What is a glycosidic bond ?

Ans. Glycosidic bond is an ester bond connecting two monosaccharides units.

Q. What are disaccharides? Give examples.

Ans. Disaccharides are carbohydrates containing two similar or dissimilar monosaccharides which are linked together by a glycosidic bond.

Disaccharides of biological significance are lactose, maltose and sucrose.

Lactose is a disaccharide where β-D-galactose is linked by β–1, 4 – linkage to β-D-glucose. Lactose is also called as milk sugar.

Maltose is a disaccharide of two glucose units which are linked together by α-1, 4-glycosidic linkage.

Sucrose consists of α-D-glucose which is linked by β – 1, 2-glycosidic linkage with β-D-fructose. It is also called as table sugar or cane sugar.

Q. Why sucrose is a nonreducing sugar ?

Ans. Since carbonyl groups (which have reducing properties) of both the monosaccharides, *i.e.* C_1 of glucose and C_2 of fructose are linked together by a glycosidic bond and there is no free carbonyl group on any of its monosaccharides, sucrose therefore is a nonreducing sugar.

On the other hand, in maltose C_1 of a glucose unit is linked to C_4 of the other glucose unit and the C_1 (the carbonyl group) of the second glucose unit is free, maltose is a reducing sugar. Similarly in lactose, C_1 of galactose is linked to C_4 of glucose and that the C_1 of glucose is free, hence lactose is also a reducing sugar.

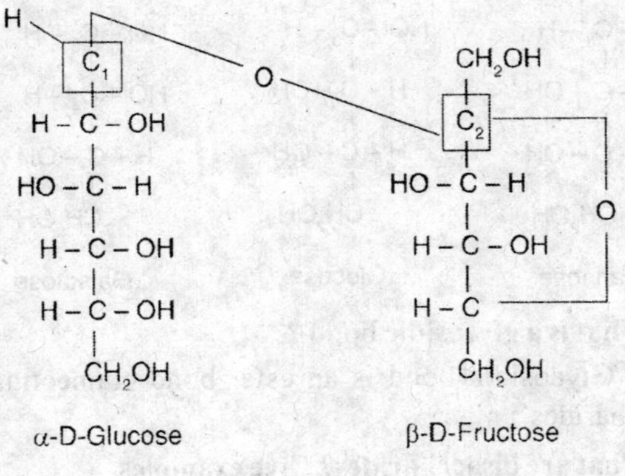

H
C_1
H – C – OH
HO – C – H
H – C – OH
H – C – OH
CH_2OH

O

CH_2OH
C_2
HO – C – H
H – C – OH
H – C
CH_2OH
O

α-D-Glucose β-D-Fructose

Sucrose

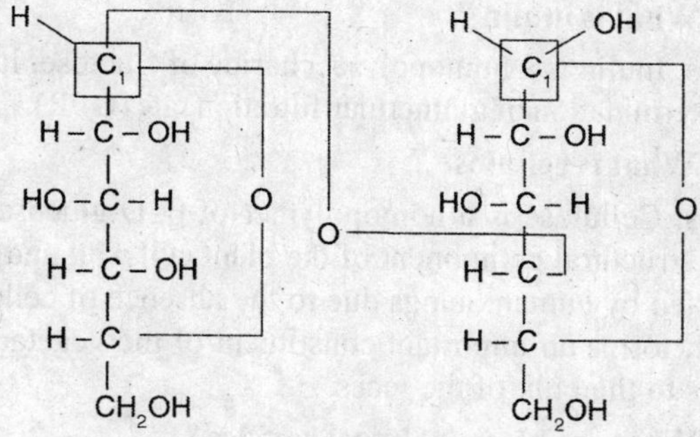

α-D-Glucose α-D-Glucose

Maltose

Q. What are polysaccharides ?

Ans. Polysaccharides are polymers containing a large number of monosaccharides which are linked together by the glycosidic bonds. When the number of monosaccharides is small, these polysaccharides are referred to as oligosaccharides.

If all the monosaccharides are of the same type, these polysaccharides are referred to as homopolysaccharides, *e.g.* starch, glycogen, etc.

On the other hand, heteropolysaccharides contain different types of monosaccharides, *e.g.* heparin, hyaluronic acid, etc.

Q. Differentiate between starch and glycogen.

Ans. Although both are homopolysaccharides of glucose, and have $\alpha - 1 \rightarrow 4$ and $\alpha - 1 \rightarrow 6$ glycosidic bonds, these two differ from each other in several ways –

Starch is a polymer of glucose found in plants. It consists of two different molecules, *i.e.* α-amylose and amylopectin. α-Amylose is a linear polymer of glucose with $\alpha - 1, 4$ -glycosidic linkage while amylopectin is a branched molecule containing both $\alpha - 1, 4$ as well as $\alpha - 1, 6$ linkages.

Starch is a main source of carbohydrate in a human diet. It gives blue color with iodine.

Glycogen is the storage form of energy in animals. It is stored in the liver and muscle. It is a highly branched structure and gives brown to red color with iodine.

Q. What is inulin ?

Ans. Inulin is a homopolysaccharide of fructose. It is used in the determination of glomerular filtration rate (GFR).

Q. What is cellulose ?

Ans. Cellulose is a homopolymer of β–D-glucose. It is a primary structural component of the plant cell wall and can not be digested by human beings due to the absence of cellulase. It however, forms an important constituent of the vegetarian diet and adds to the bulk of the feces.

Q. What are heteropolysaccharides ?

Ans. Heteropolysaccharides are also called as heteroglycans. These are the polymers of different monosaccharides or their derivatives, all of which bear negatively charged groups.

These are mainly present in association with protein in the connective tissue and are referred to as glycosaminoglycans, e.g. hyaluronic acid, chondroitin sulfates, heparin, etc.

Hyaluronic acid molecules are composed of repeating disaccharide of D-glucuronic acid and N-acetyl-D-glucosamine.

Chondroitin sulfates contain disaccharide repeating units of D-glucuronic acid and N-acetylgalactosamine sulfates.

Heparin is composed of the repeating disaccharide of D-idouronate-2-sulfate and N-sulfo-D-glucosamine -6-sulfate. Heparin inhibits clotting of blood and is widely used as anticoagulant.

Q. Describe the process of digestion of carbohydrates.

Ans. Digestion of carbohydrate starts in the mouth when it comes in contact with the saliva.

In the first step, salivary amylase hydrolyses $1\rightarrow4$ - glycosidic linkages randomly and converts starch/glycogen to various oligosaccharides and maltose. Thereafter, food reaches the duodenum where pancreatic amylase further hydrolyses these products to produce limit dextrin, maltose, isomaltose, maltotriose, and free glucose. Digestion continues in the small intestine by oligosaccharidases and various diasaccharidases, and hydrolyse oligosaccharides and disaccharides to the constituent monosaccharides.

Thus, dietary carbohydrates are hydrolyzed to monosaccharides by the combined action of salivary and pancreatic amylases, oligosaccharidases (α-dextrinases) and various disaccharidases.

Q. What is lactose intolerance ?

Ans. Lactose present in the milk is hydrolysed in the intestine by lactase to its constituent monosaccharides, *i.e.* glucose and galactose. However, some individuals lack this enzyme and can not utilize lactose. Undigested lactose is catabolized by the intestinal bacteria, mainly in the colon, and generates organic acids and CO_2. This results in irritating and painful digestive–upsets with abdominal pain (cramps), flatulence and diarrhea. Hence, due to lactase deficiency, these individuals cannot utilize milk and milk products.

Q. How are dietary carbohydrates absorbed ?

Ans. Various dietary carbohydrates are hydrolyzed to monosaccharides by the combined action of amylases, dextrinases and disaccharidases. Finally, monosaccharides, predominantly glucose, is absorbed.

Among the various monosaccarides, galactose and glucose are absorbed by the active process, fructose and mannose by facilitated diffusion while pentoses are absorbed by passive diffusion.

In the small intestine, an active sodium-dependent glucose transporter (sodium-linked glucose transporter, also referred to as SLGT-1) transports glucose from the intestinal cells.

Q. What are the pathways which utilize glucose?

Ans. Dietary glucose can be utilized in the liver and other tissues through various pathways-

(*i*) Glucose can be stored in the liver and muscle, as glycogen by the process called **glycogenesis.**

(*ii*) It is oxidized to produce energy by **glycolysis.**

(*iii*) It is also oxidized for the generation of pentoses and NADPH, *via* **HMP shunt.**

Q. What are various metabolic pathways which generate glucose ?

Ans. During hypoglycemic states, glucose can be produced in the body by the following processes–

(*i*) By breakdown of glycogen, *i.e.* **glycogenolysis.**

(*ii*) Synthesis of glucose from noncarbohydrate sources

such as amino acids, lactate or glycerol, by the process referred to as **gluconeogenesis.**

Q. What is glycolysis ?

Ans. Glycolysis is also called as Embeden - Meyerhof - pathway (Embeden - Meyerhof - Parnas pathway). It is the process of catabolism of glucose either in the presence of oxygen to pyruvic acid (mainly in the liver) or due to lack of oxygen such as during exercise in the skeletal muscle, to lactic acid.

The process where glucose is oxidized aerobically is referred to as aerobic glycolysis while the process where glucose is catabolized to lactic acid anaerobically (due to lack of oxygen) is referred to as anaerobic glycolysis.

Q. What is the difference in the action of hexokinase and glucokinase ?

Ans. Both hexokinase as well as glucokinase though act on glucose and activate it to glucose-6-phosphate, the two enzymes however, differ from each other as follows :

Hexokinase is a nonspecific enzyme and can catalyze phosphorylation of any of the hexoses. It is present in all the extrahepatic tissues. Hexokinase has high affinity and low k_m for glucose. It is allosterically inhibited by glucose-6-phosphate.

Glucokinase is an inducible enzyme. It is induced to utilize glucose in the fed state. It has low affinity and thus high k_m, for glucose.

Q. Differentiate between bisphosphate and biphosphate ?

Ans. In biphosphate two molecules of phosphoric acid are attached through each other to the same carbon atom whereas in bisphosphate two molecules of phosphoric acid are linked at two different carbon atoms.

Q. Why sodium fluoride is used for the collection of blood for glucose estimation ?

Ans. Fluoride inhibits enolase activity and thus prevents glycolysis. Therefore, fluoride (as sodium fluoride) is used as an anticoagulant during the collection of blood for glucose estimation.

Q. Give diagrammatic representation of glycolysis ?

Ans.

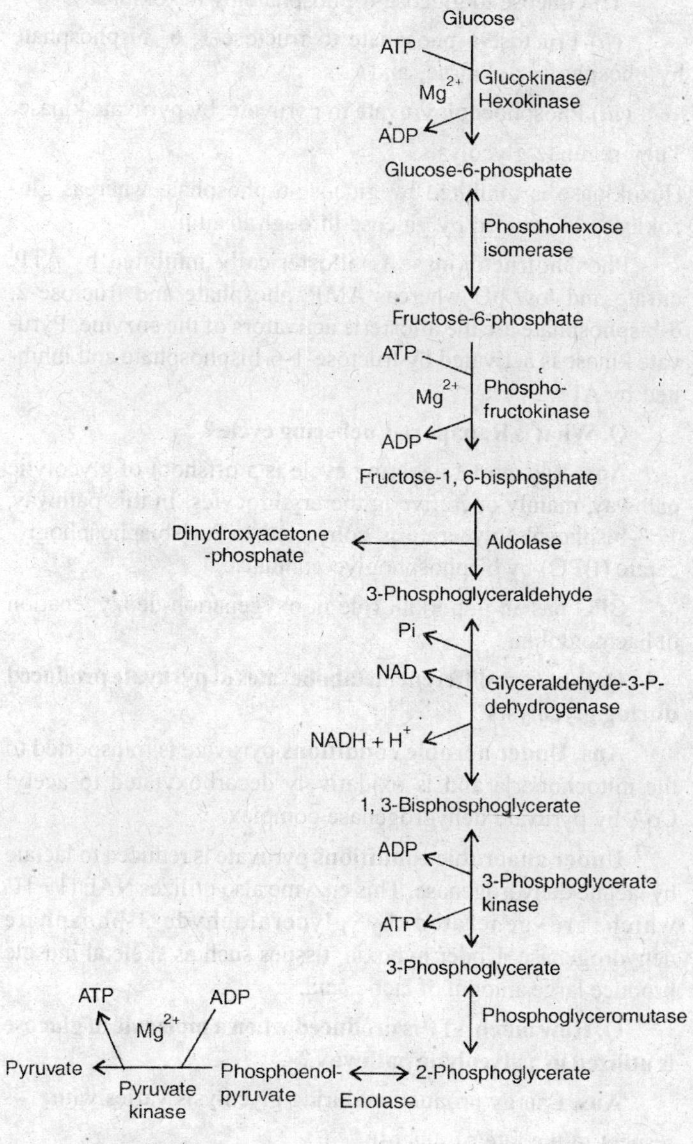

Glucose

ATP, Mg^{2+} → ADP — Glucokinase/Hexokinase

Glucose-6-phosphate

↕ Phosphohexose isomerase

Fructose-6-phosphate

ATP, Mg^{2+} → ADP — Phosphofructokinase

Fructose-1, 6-bisphosphate

↕ Aldolase → Dihydroxyacetone -phosphate

3-Phosphoglyceraldehyde

↕ Pi, NAD → NADH + H^+ — Glyceraldehyde-3-P-dehydrogenase

1, 3-Bisphosphoglycerate

↕ ADP → ATP — 3-Phosphoglycerate kinase

3-Phosphoglycerate

↕ Phosphoglyceromutase

2-Phosphoglycerate ⟷ Phosphoenol-pyruvate — Enolase

Phosphoenol-pyruvate → Pyruvate — ADP → ATP, Mg^{2+} — Pyruvate kinase

Q. How is glycolysis regulated ?

Ans. In the glycolytic pathway all the reactions are reversible except these three irreversible steps :

(*i*) Glucose to glucose-6-phosphate by hexokinase

(*ii*) Fructose-6-phosphate to fructose-1, 6- bisphosphate by phosphofructokinase, and

(*iii*) Phosphoenolpyruvate to pyruvate by pyruvate kinase.

They regulate glycolysis.

Hexokinase is inhibited by glucose-6-phosphate whereas glucokinase is induced by glucose through insulin.

Phosphofructokinase is allosterically inhibited by ATP, citrate and low pH whereas AMP, phosphate and fructose-2, 6-bisphosphate are the allosteric activators of the enzyme. Pyruvate kinase is activated by fructose-1-6-bisphosphate and inhibited by ATP.

Q. What is Rapaport-Luebering cycle ?

Ans. Rapaport-Luebering cycle is a offshoot of glycolytic pathway, mainly operative in the erythrocytes. In this pathway, 1, 3-bisphosphoglycerate is converted to 2, 3-bisphosphoglycerate (BPG) by bisphosphoglyceromutase.

BPG has an important role in oxygenation-deoxygenation of haemoglobin.

Q. What are different metabolic fates of pyruvate produced during glycolysis ?

Ans. Under aerobic conditions pyruvate is transported to the mitochondria and is oxidatively decarboxylated to acetyl CoA by pyruvate dehydrogenase complex.

Under anaerobic conditions pyruvate is reduced to lactate by lactate dehydrogenase. This enzyme also utilizes NADH + H^+ which are generated by glyceraldehyde-3-phosphate dehydrogenase. Under hypoxia, tissues such as skeletal muscle produce large amount of lactic acid.

Q. How much ATP is produced when a molecule of glucose is utilized in a glycolytic pathway ?

Ans. Energy production during glycolysis varies with respect to the fate of glucose.

Under aerobic condition.

A molecule of NADH and two molecules of ATP are generated for each molecule of a triose which is oxidized in the following three reactions :

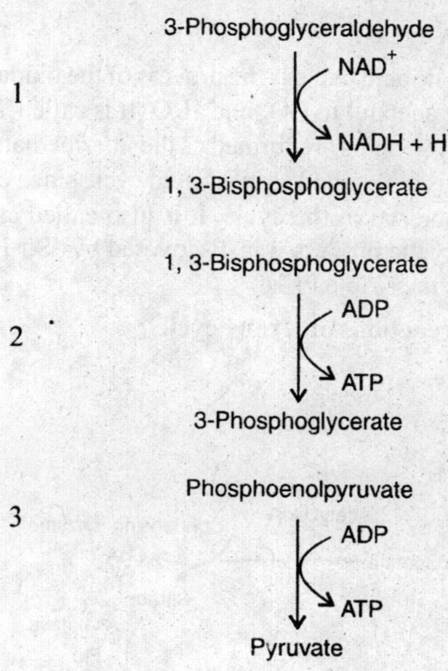

Since NADH +H$^+$ enters the mitochondrial electron transport chain and generates 3 ATP, thus a total of 5 ATP are generated per molecule of triose or 10 ATP per molecule of glucose.

Out of these 10 ATP, two are used in glycolysis in the following reactions :

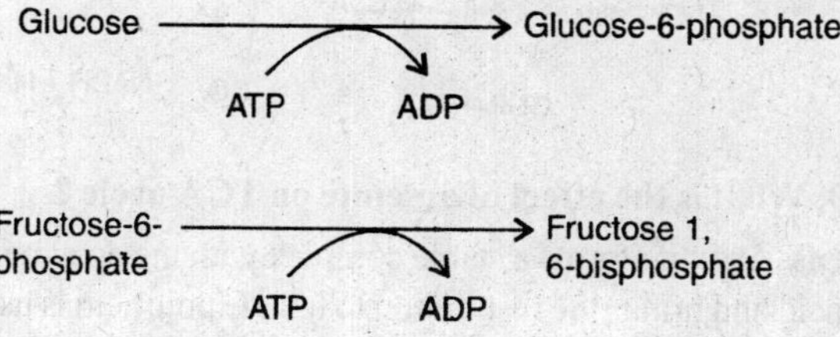

Hence, there is a net yield of 8 ATP per molecule of glucose when it is oxidized to pyruvate, aerobically.

Under anaerobic conditions.

Since NADH $+H^+$ is reutilized, only two ATP are produced per triose or 4 ATP are produced for each glucose molecule.

Q. What is TCA cycle ?

Ans. Tricarboxylic acid cycle is the process of the oxidation of acetyl CoA (active acetate) to CO_2 and H_2O. It is called TCA cycle, since tricarboxylic acids are formed as the intermediates in the cycle. TCA cycle is also called as citric acid cycle since citric acid is the starting material of the cycle. It is also called as the Krebs cycle, as this cyclic process was discovered by "Sir Hans Krebs", a recipient of the Noble Prize.

Q. Outline the reactions of Krebs cycle ?

Ans.

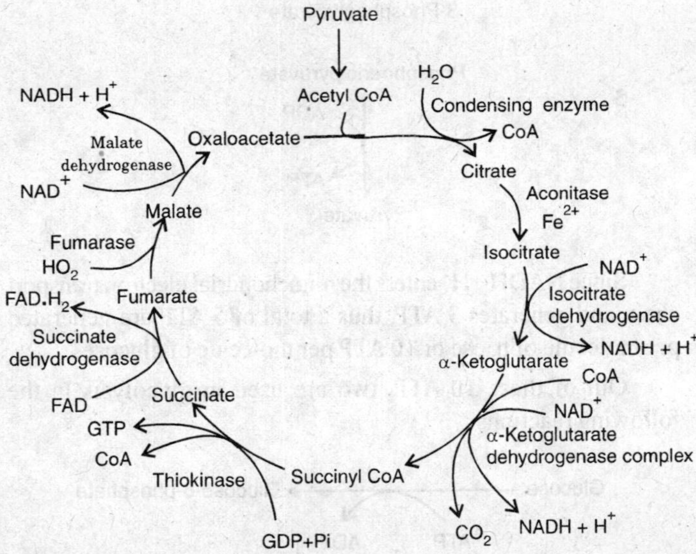

Q. What is the effect of arsenite on TCA cycle ?

Ans. Arsenite forms a stable complex with the enzyme bound lipoic acid and brings the respiration to halt. Lipoic acid is used as a coenzyme with pyruvate dehydrogenase complex in the TCA cycle.

Q. How does malonate acts as inhibitor in TCA cycle?

Ans. Malonate, due to structural similarity to succinate, competitively inhibits succinate dehydrogenase.

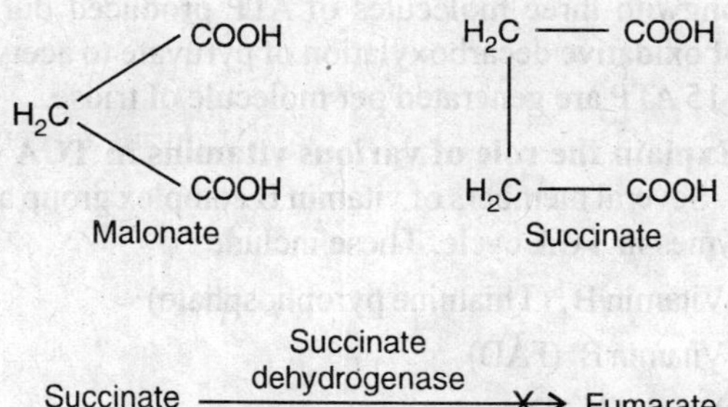

Q. How much ATP is produced during TCA cycle ?

Ans. 1. Conversion of pyruvate to acetyl CoA generates NADH + H^+ which produces three ATP.

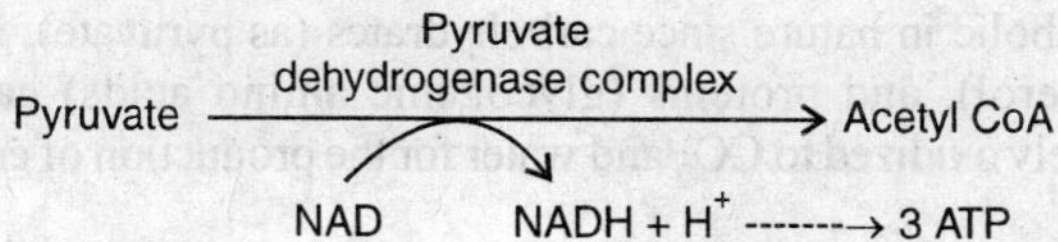

2. Acetyl CoA in turn is utilized in the TCA cycle and is converted to CO_2 and H_2O. During this process three NADH + H^+ and one $FADH_2$ are generated. This in turn yields 3x3+2 = 11 ATP.

3. Their is also a production of one ATP by substrate level phosphorylation, during the conversion of succinyl CoA to succinate.

Thus, a total of 12 ATP are generated per molecule of acetyl CoA.

Reaction	**Reducing equivalents produced**	**Number of ATP produced**
Isocitrate → α-Ketoglutarate	NADH+H^+	3
α-Ketoglutarate → Succinyl CoA	NADH+H^+	3
Succinyl CoA → Succinate (Substrate level phosphorylation)	—	1
Succinate → Fumarate	$FAD.H_2$	2
Malate → oxaloacetate	NADH+H^+	3
Total		**12**

Alongwith three molecules of ATP produced during the process of oxidative decarboxylation of pyruvate to acetyl CoA, a total of 15 ATP are generated per molecule of triose.

Q. Explain the role of various vitamins in TCA cycle ?

Ans. Several members of vitamin B complex group are used as coenzymes in TCA cycle. These include :

(*i*) Vitamin B_1 (Thiamine pyrophosphate)

(*ii*) Vitamin B_2 (FAD)

(*iii*) Niacin (NAD^+)

(*iv*) Pantothenic acid (as a constituent of coenzyme A), and

(*v*) Lipoic acid

Q. What is the biological significance of TCA cycle ?

Ans. TCA cycle is Amphibolic in nature as it has a dual role. It is important, both in the oxidative as well as synthetic processes. It is catabolic in nature since carbohydrates (as pyruvate), lipids (as glycerol), and proteins (glycogenic amino acids) can be completely oxidized to CO_2 and water for the production of energy (ATP).

It is also an anabolic process since various intermediates of the cycle (α-keto acids) can be used for the synthesis of nonessential amino acids, e.g. oxaloacetate to aspartate.

Q. How is TCA cycle regulated ?

Ans. TCA cycle is regulated at the level of citrate synthase, isocitrate dehydrogenase and α-ketoglutarate dehydrogenase which are the rate limiting enzymes of this cycle.

The most important regulators of the cycle are acetyl CoA, oxaloacetate and NADH.

Q. What is Gluconeogenesis ?

Ans. Gluconeogenesis, also called as neoglucogenesis, is the process of synthesis of glucose from various sources other than carbohydrates, *e.g.* –

glucogenic amino acids,

lactate,

glycerol, and

propionyl CoA.

Gluconeogenesis occurs during prolonged fasting or on a low carbohydrate diet, to maintain blood glucose level.

Q. How glucogenic amino acids are converted to glucose ?

Ans. Glucogenic amino acids form certain intermediates of the TCA cycle and in turn can be converted to glucose *via* pyruvate, e.g. alanine.

Q. How is lactate converted to glucose ?

Ans. Lactate is converted to pyruvate by lactate dehydrogenase. Pyruvate in turn can form glucose by the reversal of glycolysis.

$$\text{Lactate} \xrightarrow{\text{Lactate dehydrogenase}} \text{Pyruvate} \dashrightarrow \text{Glucose}$$

Q. How is glycerol converted to glucose ?

Ans. Glycerol is produced as a result of lipolysis in the adipose tissue. It is converted to glycerolphosphate, in the liver by the enzyme glycerol kinase. Glycerolphosphate is subsequently oxidized by a dehydrogenase to dihydroxyacetonephosphate which subsquently can enter glycolysis.

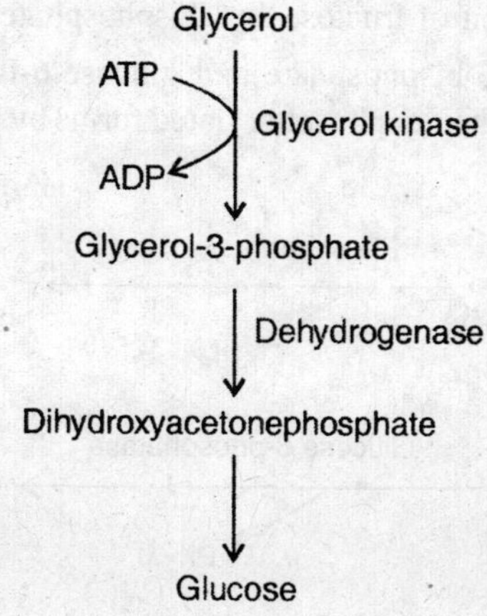

Q. How glycolysis can be reversed ?

Ans. Glycolysis utilizes glucose whereas reversal of the process produces glucose.

Glycolysis is a reversible process except the following three reaction :

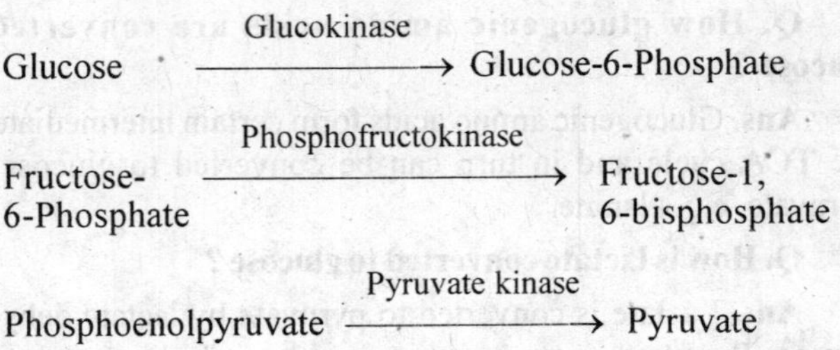

These three irreversible steps, catalyzed by different kinases, can be reversed during the process of gluconeogenesis, as follows :

Pyruvate is first converted to oxaloacetate by pyruvate carboxylase which is subsequently converted to phosphoenol-pyruvate by carboxykinase.

$$\text{Pyruvate} \xrightarrow{\text{Pyruvate carboxylase}} \text{Oxaloacetate} \xrightarrow{\text{Carboxykinase}} \text{Phosphoenolpyruvate}$$

Subsequent steps in the glycolysis thereafter are reversible upto the generation of fructose-1, 6-bisphosphate.

Fructose-1, 6-bisphosphate and glucose-6-Phosphate can be hydrolyzed to their dephosphorylated forms by the respective phosphatases :

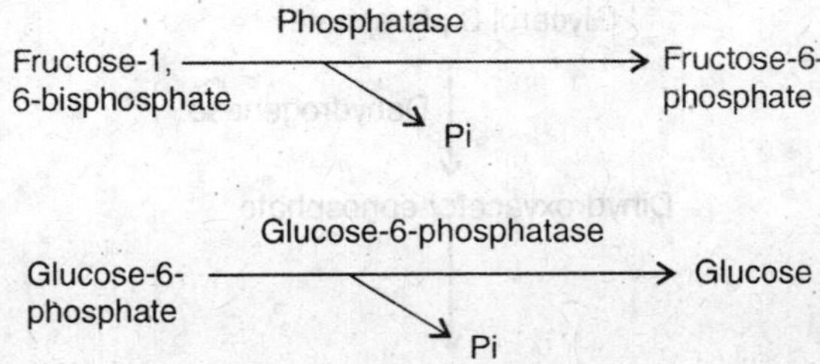

Q. What is Cori cycle ?

Ans. Cori cycle is the process of utilization of glycogen in the muscle, generating lactic acid which is transported to the liver, for gluconeogenesis. Glucose so generated is transported back to the muscle for its reutilization as glycogen.

Q. What is glucose-alanine cycle ?

Ans. During starvation, protein breakdown in muscle results in the release of various amino acids, amongst which alanine predominates. It is transported to the liver and converted to pyruvate which can be used for the synthesis of glucose by the process of gluconeogenesis and thus helps in maintaining blood glucose level.

During glycolysis, glucose is changed to pyruvate which can be converted to alanine by transamination. This interconversion of glucose to alanine and vice-versa is referred to as glucose-alanine cycle.

Q. How is gluconeogenesis regulated ?

Ans. Gluconeogenesis is the process of the reversal of glycolysis. The two processes are regulated, reciprocally, to maintain blood glucose level.

Regulatory enzymes of glycolysis, *i.e.* hexokinase, phosphofructokinase and pyruvate kinase catalyze irreversible steps. Reversal of these reactions takes place by pyruvate carboxylase, phosphoenolpyruvate carboxykinase, fructose-1, 6 bisphos-phatase and glucose-6-phosphatase. These are the regulatory enzymes of gluconeogenesis.

Glucagon and insulin regulate the two processes by inducing activation/inactivation of the enzymes of the corresponding pathways.

Q. What is Glycogen ?

Ans. Glycogen is the storage form of glucose, present in several tissues mainly in the liver and muscle in the body. It is a highly branched polymer of α-D-glucose. Glucose residues are linked together by α-1, 4 - and α-1, 6-glycosidic linkages.

Q. What is the significance of glycogen which is stored in the liver and that in the muscle ?

Ans. Glycogen stored in the liver is converted to free

glucose and helps to maintain fasting blood glucose level whereas glycogen stored in the muscle is used as a source of energy during muscular contraction.

Muscle glycogen can not be converted to free glucose due to the absence of glucose-6-phosphatase. Hence, muscle glycogen can not be used for the maintenance of blood glucose level.

Q. What is glycogenesis ?

Ans. Glycogenesis is the process of the synthesis of glycogen from glucose. It occurs during the fed state.

Q. Out line the steps of glycogenesis.

Ans.

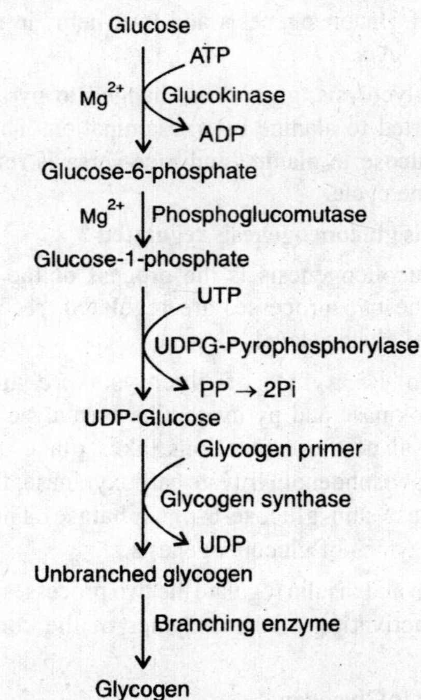

Q. What is the regulatory enzyme of glycogenesis ?

Ans. Glycogen synthase is a key enzyme of glycogenesis. The enzyme exists in two forms called glycogen synthase ***a*** and glycogen synthase ***b***. Glycogen synthase a is the dephosphorylated form of the enzyme and is physiologically active. Glycogen synthase ***b*** is the phosphorylated form of the enzyme and is physiologically inactive.

The enzyme is under allosteric control. Dephosphorylated form of the enzyme (glycogen synthase a or active enzyme) can be inactivated in the presence of ATP. It is phosphorylated by cyclic AMP dependent protein kinase.

Inactive enzyme (phosphorylated synthase b) is converted to active (dephosphorylated synthase a) by protein phosphatase 1.

Q. What is glycogenolysis ?

Ans. Glycogenolysis is the process of the breakdown of glycogen either to free glucose in the liver or to glucose-6-PO_4 and finally to lactate, in the muscle.

Q. Describe the process of glycogenolysis ?

Ans. Glycogenolysis occurs as follows :

1. Sequentially, glycogen is hydrolyzed and releases glucose-1-phosphate. This reaction is catalyzed by phosphorylase.

Phosphorylase removes glucose units leaving four glucose residues near the branch point.

2. Out of the four glucose units near the branch, three glucose units are transferred onto the another nearby branch. This reaction is catalyzed by glycosyltransferase.

The remaining glucose unit, at the branch point, is removed as glucose by amylo-1, 6-glucosidase.

Glucosyltransferase and amylo-1, 6-glucosidase are two subunits of the debranching enzyme.

3. Glucose-1-phosphate is subsquently converted to glucose-6-phosphate. This reaction is catalyzed by phosphoglucomutase.

Glucose-6-phosphate is the end product of glycogenolysis in the muscle. Since, muscle lacks glucose-6-phosphatase, glucose-6-phosphate enters anaerobic glycolysis and is converted to lactic acid during excercise.

4. In the liver, glucose-6-phosphate is further hydrolyzed to glucose. Since liver contains glucose-6-phosphatase, liver glycogen is used to maintain blood glucose level during fasting.

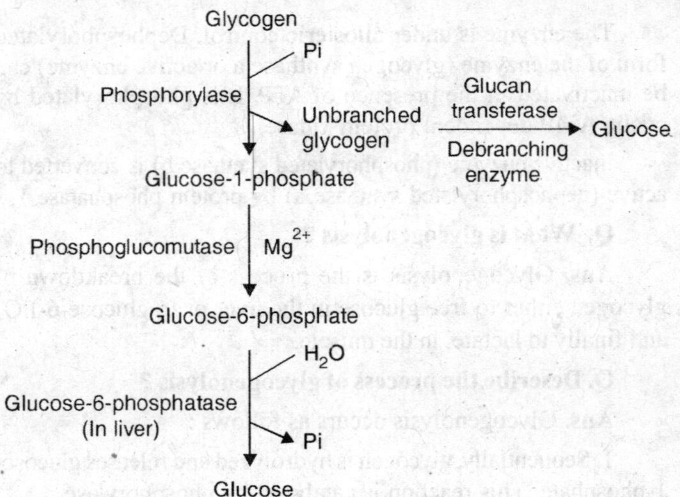

Q. Describe various glycogen storage diseases ?

Ans. Glycogen storage diseases include metabolic disorders of glycogen, resulting in abnormal quantity or quality of glycogen in different tissues. These include :

1. Glycogen storage disease type I. It is also known as Von Geirke's disease and is due to the deficiency of glucose-6-phosphatase in the liver. It is characterized by hypoglycemia, lactic acidosis and hyperuricemia. Large amount of glycogen is stored in the liver and the kidney.

2. Glycogen storage disease type II. It is also referred to as Pompe's disease. It is due to the deficiency of lysosomal acid maltase, resulting in the accumulation of glycogen in the lysosomes.

3. Glycogen storage disease type III. It is also referred to as limit dextrinosis or Forbe's disease or Cori's disease. It is due to the deficiency of the debranching enzyme and is characterized by the accumulation of glycogen which is partially hydrolyzed to limit dextrin, and accumulates in the liver, muscle and heart.

4. Glycogen storage disease type IV. This is also referred to as amylopectinosis or Anderson's disease. It is due to the deficiency of the branching enzyme and results in the accumulation of unbranched glycogen.

5. Glycogen storage disease type V. This is referred to as McArdle's syndrome. It is due to the deficiency of muscle phosphorylase and is characterized by the accumulation of glycogen in the muscle. There is dimished exercise tolerance.

6. Glycogen storage disease type VI. It is also refered to as Her's disease. It is due to the deficiency of liver phosphorylase and results in glycogen deposition in the liver.

Q. Outline HMP shunt.

Ans.

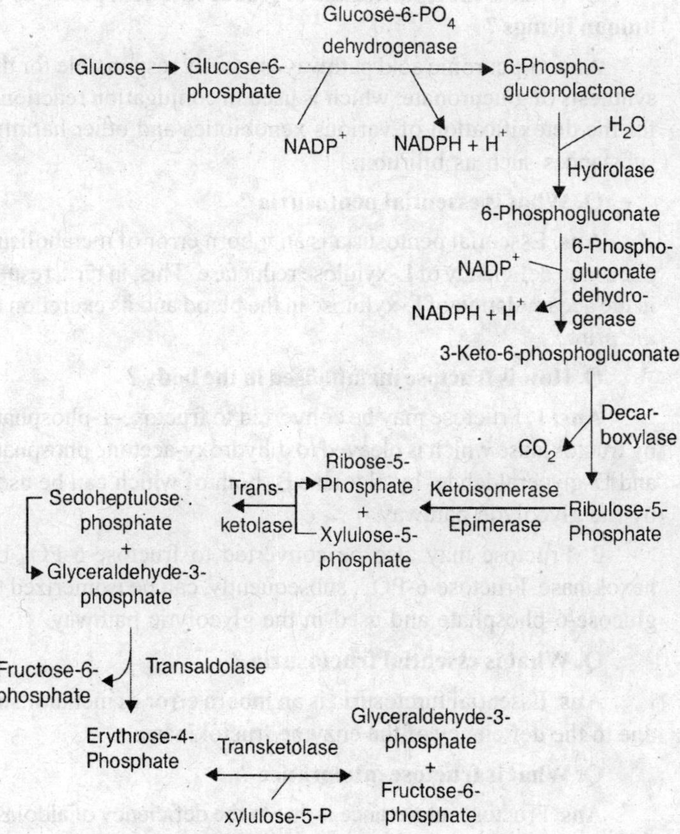

Q. Describe biological significance of the hexose monophosphate shunt.

Ans. Hexose monophosphate shunt (HMP shunt) is also referred to as pentose phosphate pathway. It is the second major pathway for glucose metabolism, as an alternative to glycolysis.

This pathway provides pentoses (ribose) required for the biosynthesis of nucleic acids. It also generates NADPH which is required for various anabolic reactins, such as in the biosynthesis of lipids and steroids. It is also important for the detoxification of H_2O_2, particularly within the erythrocytes and prevent haemolysis.

Q. What is the significance of glucuronic acid pathway in human beings ?

Ans. Glucuronic acid pathway is mainly responsible for the synthesis of glucuronate, which is used in conjugation reactions, for the detoxification of various xenobiotics and other harmful substances such as bilirubin.

Q. What is essential pentosuria ?

Ans. Essential pentosuria is an inborn error of metabolism, due to the deficiency of L-xylulose reductase. This, in turn, results in the accumulation of L-xylulose in the blood and its excretion in the urine.

Q. How is fructose metabolised in the body ?

Ans. 1. Fructose may be converted to fructose-1-phosphate by fructokinase which is cleaved to dihydroxy-acetone phosphate and D-glyceraldehde, by aldolase B, both of which can be used by the glycolytic pathway.

2. Fructose may also be converted to fructose-6-PO_4 by hexokinase. Fructose-6-PO_4, subsequently, can be isomerized to glucose-6-phosphate and used in the glycolytic pathway.

Q. What is essential fructosuria ?

Ans. Essential furctosuria is an inborn error of metabolism, due to the deficiency of the enzyme fructokinase.

Q. What is fructose intolerance ?

Ans. Fructose intolerance is due to the deficiency of aldolase B. Such individuals cannot tolerate high fructose and/or sucrose diet.

Q. How galactose is metabolized in the body ?

Ans. Galactose is converted to galactose-1-PO_4, by galactokinase. Galactose-1-PO_4 is then changed to UDP-galactose by galactose-1-phosphate uridyltransferase. Subsequently, UDP-galactose is utilized differently in different tissues –

1. In the liver, UDP-galactose is changed to UDP-glucose by epimerase. UDP-glucose, in turn, may be used in glycogen synthesis.

2. In the brain, UDP-galactose is utilized for the synthesis of sphingomyelins and cerebrosides.

3. In the lactating mammary glands, it is used in the synthesis of lactose.

Q. What is galactosemia ?

Ans. Galactosemia is an inborn error of galactose metabolism due to the deficiency of the enzyme galactose-1-phosphate uridyltransferase which results in the accumulation of galactose in the blood, eye lens, and galactosuria. Galactosemia also results in mental retardation and cataract formation.

Q. Describe hormonal regulation of blood glucose ?

Ans. Several hormones regulate blood glucose concentraiton –

Insulin – It lowers blood glucose concentration by various mechanisms –

(*i*) Increases uptake by the liver

(*ii*) Promotes glycolysis

(*iii*) Stimulates glycogenesis

(*iv*) Inhibits glycogenolysis and gluconeogenesis

(*v*) Promotes transport of amino acids and protein biosynthesis

(*vi*) Inhibits ketogenesis

(*vii*) Promotes lipogenesis.

Glucagon – Glucagon promotes glycogenolysis and gluconeogenesis in the liver.

Epinephrine –

(*i*) Epinephrine promotes glycogenolysis, both in the liver and muscle

(*ii*) It also promotes gluconeogenesis in the liver.

(*iii*) It stimulates release of glucagon and inhibits insulin secretion.

Glucocorticoids –

(*i*) Glucocorticoids stimulate protein metabolism in the muscle

(*ii*) They also stimulate lipolysis

(*iii*) They promote gluconeogenesis

(*iv*) They inhibit glucose oxidation

Growth hormone –

(*i*) Growth hormone decreases glucose uptake and its utilization by the muscle.

(*ii*) It promotes lipolysis in the adipose tissue.

Thyroid hormones –

(*i*) They stimulate intestinal absorption of glucose

(*ii*) They promote glycogenolysis and gluconeogenesis in the liver.

Q. What is diabetes mellitus ?

Ans. Diabetes mellitus is a state of chronic hyperglycemia due to genetic and/or environmental factors. Though it is a multifactorial disease. Most common cause include deficiency of insulin. It may be of two types :

1. Insulin – dependent diabetes mellitus
2. Non-insulin dependent diabetes mellitus.

Diabetes mellitus, in general, is characterized by polyuria, polydypsia and polyphagia. Long term complicaitons of the disease include nephropathy, retinopathy, neuropathy, cardiopathy, infections and delayed wound healing.

Q. Differentiate between insulin-dependent diabetes mellitus (IDDM) and non-insulin dependent diabetes mellitus (NIDDM).

Ans.

Insulin-dependent diabetes mellitus	Non-insulin dependent diabetes mellitus
(*i*) It is referred to as type-I or juvenile onset diabetes since it is prevalent in mainly youger age group (below 40 yrs)	It is referred to as type-II or maturity onset diabetes and prevalent mainly in the older age group (more than 40 yrs)

(*ii*) It is present in nearly 20% of population	It is present in 80% of the population
(*iii*) It is characterized by hyperglycemia, hyperlipoproteinemia (mainly chylomicrons and VLDL) and ketoacidosis.	It is characterized by hyperglycemia and hypertriglyceridemia (VLDL)
(*iv*) Their is a weight loss	These individuals are generally obese.
(*v*) Plasma insulin is low to absent due to destruction of β-cells and patient responds to insulin therapy.	Plasma insulin is normal to high. They are insulin resistant and respond to oral hypoglycemic agents.

5

Lipids

Q. What are lipids ?

Ans. Lipids are a heterogeneous group of hydrophobic organic molecules found in the cell. They are sparingly soluble in water but soluble in organic solvents, such as alcohol.

Q. Classify lipids.

Ans. Lipids are classified into 3 groups :

(*i*) **Simple lipids :** These are the esters of fatty acids with alcohol, *e.g.* triacylglycerols (triglycerides).

(*ii*) **Compound lipids :** In addition to fatty acids and alcohol, these lipids also contain some amphipathic group, *e.g.* phospholipids or proteolipids (lipoproteins), etc.

(*iii*) **Derived lipids :** These are the hydrolyzed products of simple and compound lipids, *e.g.* fatty acids, glycerol, etc.

Q. What is the physiological significance of lipids?

Ans. Lipids are important for the body, as :

(*i*) Lipids have high calorific value, *i.e.* 9 Kcal/g, as compared to 4 Kcal obtained from 1 g of carbohydrates or proteins.

(*ii*) Fat, present in the subcutaneous tissue, acts as an insulating material.

(*iii*) Lipids are the important component of biological membrane.

(*iv*) Dietary lipids are the source of fat soluble vitamins and essential fatty acids.

(*v*) Lipids are also involved in intercellular as well as intracellular signalling, *e.g.* 1, 2-diacylglycerol and inositol-phosphate.

Q. What are fatty acids ?

Ans. Fatty acids are carboxylic acids with a hydrocarbon side chain which may be saturated or unsaturated.

Q. What is the delta system of nomenclature of fatty acids?

Ans. Various carbon atoms in a fatty acid are numbered,

beginning with the carboxyl carbon as the first carbon. The next carbon atom is designated as carbon number 2, which is also referred to as α-carbon atom. Accordingly, carbon atom 3 is referred to as β-carbon atom, and so on. This system of nomenclature is called as delta system.

$$\underset{}{CH_3}-CH_2-CH_2-CH_2-CH_2-CH_2-\underset{4}{\overset{\gamma}{CH_2}}-\underset{3}{\overset{\beta}{CH_2}}-\underset{2}{\overset{\alpha}{CH_2}}-\underset{1}{COOH}$$

Q. What is the omega system of nomenclature of a fatty acid ?

Ans. In this system of nomenclature, carbon atoms are counted from the methyl (CH_3 –) terminal end where a double bond first appears. This carbon atom at the terminal methyl group is referred to as omega carbon atom.

$$\underset{1}{\overset{\omega}{CH_3}}-(CH_2)_4-\underset{6}{CH}=CH-CH_2-CH=CH-(CH_2)_7-COOH$$

(An ω–6 fatty acid)

Q. What are essential fatty acids ?

Ans. Two polyunsaturated fatty acids, *i.e.* linoleic acid and linolenic acid are collectively referred to as essential fatty acids because they cannot be synthesized in the body and are essentially required in the diet. Their deficiency results in growth retardation, hair loss, dermatitis and impaired wound healing. If there is a lack of linolenic acid in the diet, even arachidonic acid becomes dietary essential.

Q. What is the significance of essential fatty acids ?

Ans. These are essential for the body, since –

(*i*) They help in the esterification of free cholesterol and thus lower serum cholesterol level.

(*ii*) One of these essential fatty acids, *i.e.* arachidonic acid (with 20 C and 4 double bonds) is a precursor of important group of compounds referred to as eicosanoids, such as prostaglandins, thromboxanes and leukotrienes.

(*iii*) Deficiency of these fatty acids leads to growth retardation and alter membrane fragility.

Q. What are eicosanoids ?

Ans. Eicosanoids are 20 carbon compounds such as prostaglandins, thromboxanes and leukotrienes which have various physiological and pharmacological actions. These are derived from arachidonic acid and usually have three to five double bonds.

Q. What are prostaglandins ?

Ans. Prostaglandins are a group of 20 carbon compounds derived from arachidonic acid. They are named, because of their probable origin from the prostate gland. Different prostaglandins may have different number of double bonds, hydroxyl groups and/or a keto group, and a cyclopentane ring,e.g. PGE_2.

Q. What is the physiological significance of prostaglandins ?

Ans. Prostaglandins are produced and released by all mammalian cells and tissues except RBC, and have various physiological and pharmacological actions (as local hormones). They have effects on smooth muscle, especially promote contraction of the intestinal and uterine muscles, and lower blood pressure. PGE_2 is a major prostaglandin produced in the kidney and spleen while PGI_2 is produced in blood vessels.

Q. What are prostacyclins ?

Ans. Prostacylins are PGI_2 type of prostaglandins which are synthesized in the vascular endothelium. They have powerful vasodilatory action and inhibit platelet aggregation.

Q. What is the effect of anti-inflammatory drugs on prostaglandin synthesis ?

Ans. Nonsteroidal anti-inflammatory drugs such as aspirin and indomethacin inhibit cyclo-oxygenase 1 activity and thus block the synthesis of prostaglandins.

On the other hand, steroidal anti-inflammatory drugs like prednisolone and hydrocortisone inhibit the release of prostaglandins.

Q. What are thromboxanes ?

Ans. Thromboxanes are highly active metabolites of PGG_2 and PGH_2 mainly formed in platelets. Because they have thrombus forming potential, hence called thromboxanes. They cause release of serotonin and calcium, from the platelet granules, produce vasoconstriction, increase blood pressure, cause contraction of the arterial smooth muscle and platelet aggregation.

Q. What are leukotrienes ?

Ans. Leukotrienes are a group of compounds synthesized in the leucocytes, mast cells and macrophages, in response to immunological as well as noninflammatory stimuli.

They stimulate mucus secretion and cause constriction of bronchial muscles. Elevated levels of leukotrienes have been shown to be associated with various inflammatory and hypersensitive disorders like asthma.

Q. What are triacylglycerols ?

Ans. Triacylglycerols (triglycerides) are fatty acid esters of glycerol. These are also called as neutral fats.

Glycerol with one molecule of a fatty acid is called monoacylglycerol, with two molecules diacylglycerol and with three molecules as triacylglycerol.

A molecule of triacylglycerol may contain 3 similar or dissimilar fatty acids which may be saturated or unsaturated. Generally, an unsaturated fatty acid is present at C2.

Q. What is rancidity ?

Ans. Rancidity is the development of unpleasant odour and taste due to auto-oxidation of fatty acids, particularly, during partial hydrolysis of triacylglycerols. Antioxidants such as vitamin E prevent this process.

Q. What is saponification ?

Ans. Saponification is defined as the process of hydrolysis of fat or oil, by alkali.

Q. What is saponification number ?

Ans. Saponification number is the number of mg of KOH required to saponify 1g of fat or oil.

Q. What is iodine number ?

Ans. Iodine number is defined as the grams of iodine that are used for the iodination (halogenation) of 100g of fat. Since iodine is taken up at the double bond, iodine number thus is a measure of the degree of unsaturation of fat.

Q. What are phospholipids ?

Ans. Phospholipids are the heterogenous group of compounds which in addition to long chain fatty acids and alcohol, contain phosphoric acid and a polar group such as a nitrogenous base, an amino acid or polyalcohol like inositol. Along- with cholesterol, phospholipids form important constituents of the cell membrane.

Q. What is the significance of 1, 2 diacylglycerol ?

Ans. 1, 2-diacylglycerol is a hydrolyzed product of glycerophospholipid, by the action of phospholipase C. It acts as an intra-cellular signal transducer and activates protein kinase C to stimulate a hormonal action.

Q. What are lecithins ?

Ans. Lecithins are phospholipids containing choline (a nitrogenous base) which is attached at C3. These are the surface active agents and help in the emulsification of fat. These are widely distributed in the brain, nerve cells, sperm and egg yolk.

Q. What are cephalins ?

Ans. Cephalins are glycerophospholipids which contain ethanolamine or serine as a nitrogenous base which is attached to phosphoric acid at C3. Cephalins are present in the erythrocyte membrane and brain tissue.

Q. What are plasmalogens ?

Ans. Plasmalogens are glycerophospholipids in which C–1 of glycerol is linked to an unsaturated fatty acid; by ether linkage. These are normally present in the cardiac muscle, skeletal muscle and semen. Because of their high concentration in the cardiac muscle, plasmalogens are also referred to as cardiolipins.

Q. What are ceramides ?

Ans. Ceramides are the N-acetyl derivatives of the unsaturated aminoalcohol referred to as sphingosine. These are the parent compounds of sphingolipids.

Q. What are sphingolipids ?

Ans. Sphingolipids are phospholipids which contain an unsaturated amino alcohol, sphingosine.

Sphingomyelins, cerebrosides and gangliosides are some of the sphingolipids that are present in high concentration, in the myelin sheath.

Q. What is the physiological significance of lecithin : sphingomyelin ratio ?

Ans. Lecithin : sphingomyelin ratio (L/S ratio), in the amniotic fluid, is used in the evaluation of fetal lung maturity.

At term, there is a marked increase in lecithin concentration and L/S ratio increases to more than 5. In a premature low birth weight fetus, the ratio is less than two. Such infants develop respiratory distress syndrome.

Q. What are cerebrosides ?

Ans. Cerebrosides are glycosphingolipids. In addition to sphingosine and a nitrogenous base, they also contain a sugar residue such as galactose or glucose. Common cerebrosides include galactocerebrosides and glucocerebrosides. These are found in high concentration in the white matter of the brain and myelin sheath of the nerves.

Q. What are gangliosides ?

Ans. Gangliosides are the oligosaccharide derivatives of ceramides. The oligosaccharide unit includes hexosamines (glucosamine or galactosamine) and atleast one molecule of N-acetylneuraminic acid (NANA, also called as sialic acid). These are primarily the components of the cell membrane and the brain lipids, *e.g.* GM_1, GM_2 , etc.

Q. What is cholesterol ?

Ans. Cholesterol is a steroid alcohol (sterol) found in animals. It has 27 carbon atoms, a hydroxyl group at C3, a double bond between C5 and C6, two methyl groups at C18 and C19, and a side chain at C17.

The basic steroid ring is called as cyclopentanoperhydrophenanthrene ring.

Normal serum cholesterol level is 130-200 mg/dL.

Q. What is the significance of cholesterol for our body ?

Ans. Cholesterol is a metabolic precursor of steroid hormones, vitamin D and bile acids/salts.

Q. What are lipoproteins ?

Ans. Lipoproteins are compound lipids where lipids are associated with proteins. These are also called as proteolipids.

The protein part of a lipoprotein is called apolipoprotein or apoprotein.

Q. Classify lipoproteins ?

Ans. Lipoproteins are classified as :

(*i*) Chylomicrons,

(*ii*) Very low density lipoproteins (VLDL),

(*iii*) Low density lipoproteins (LDL), and

(*iv*) High density lipoproteins (HDL).

Q. Describe the process of digestion of lipids ?

Ans. Digestion of lipids is initiated by an acid stable lipase which is present in sublignual glands as well as stomach. However, main enzyme for the hydrolysis of triacylglycerols is pancreatic lipase which is present in the pancreatic juice. It is also referred to as steapsin. The main site of digestion of lipid is the small intestine.

Pancreatic lipase attacks long chain fatty acids (with more than ten carbon atoms). This enzyme also requires co-lipase. Pancreatic lipase is an α-lipase which hydrolyzes fatty acids from the first and the third position of a triglyceride and results in the formation of β-monoacylglycerols and free fatty acids. These are incorporated into the micelles which provide vehicle for transport of lipids from the intestinal lumen.

Pancreatic juice also contains an esterase which hydrolyzes cholesterol esters, esters of retenoic acid and monoacylglycerols. Pancreatic juice is also rich in phospholipases which hydrolyze phospholipids.

Various dietary lipids are incorporated into micelles which also requires bile acids that help in the absorption of dietary lipids.

Q. Explain the process of the absorption of the digestion products of lipids.

Ans. Absorption of dietary lipids mainly free fatty acids and β-monoacylglycerols occurs from micelles in the brush border of the epithelial cells of small intestine, by passive diffusion.

Within the intestinal cells, the fate of absorbed fatty acids depend upon their chain length. Short to medium chain fatty acids (with less than ten carbon atoms) as well as unsaturated fatty acids are absorbed more readily and pass through the cell without any modification, and reach the liver directly via the portal blood. On the other hand, long chain fatty acids (with more than 12 carbons) bye-pass the liver and are transported to the endoplasmic reticulum, where they are used for the resynthesis of triacylglycerols. Within the intestinal mucosa, cholesterol is resterified. The hydrolyzed products of phospholipids, *i.e.*, phosphoric acid, nitrogenous bases, glycerol and fatty acids are also utilized for the resynthesis of phospholipids. Triacylglycerols, which are resynthesized in the intestinal mucosa form fat globules to which apoproteins, and phospholipids are also added. These fat globules are released from the intestine via lymph vessels as milky juice called **chyle** and are several microns in diameter. Hence these particles are called chylomicrons. Within the membrane bound vesicles, chylomicrons migrate to the Golgi apparatus through basolateral membrane and are released into the intercellular space. Intestinal lymph vessels, via the thoracic duct, drain into the large veins and thus deliver the dietary lipids to the peripheral tissues. Most of these triacylglycerols are taken up by the cells of tissues like adipose tissue and muscles for storage and metabolism, respectively, and bye-pass the liver.

Q. Explain the process of β-oxidation of fatty acids.

Ans. The process of the oxidation of a fatty acid is referred to as β-oxidation since during this process a fattyacyl chain is cleaved at the β-carbon atom producing acetyl CoA. The process of β-oxidation occurs in the following steps:

1. Activation of a fatty acid. During this process, a fatty acid is converted to fattyacyl CoA, by the enzyme fattyacyl CoA synthetase.

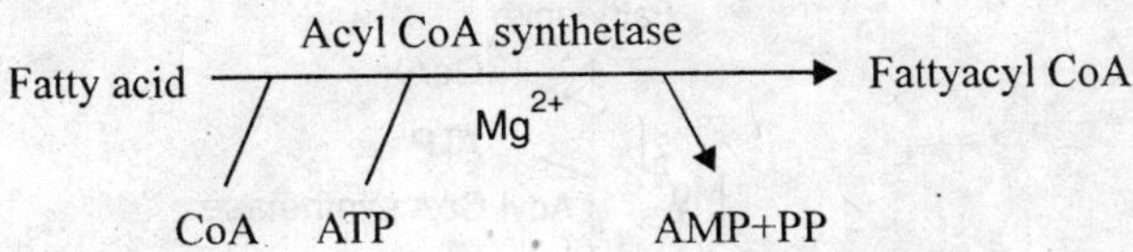

Acyl CoA synthetase is a group of enzymes, also referred to as thiokinases. Different types of thiokinases are found, each one of which is specific for a type of fatty acid, depending upon its chain length. Activation of a fatty acid takes place in the cytoplasm.

2. Transport of the fatty acid across the mitochondrial

membrane. Fattyacyl CoA is formed in the cytoplasm but it cannot penetrate the inner mitochondrial membrane. The transport of a fatty acid is facilitated by carnitine with the help of carnitine palmitoyltransferase.

Carnitine palmitoyltransferase-I (CPT-I) helps in the formation of acylcarnitine. Translocase and CPT-II helps in translocation of the acyl group into the mitochondria as fattyacyl CoA.

3. The process of β-oxidation. Fatty acyl CoA, after its entry into mitochondria, undergoes β-oxidation as follows :

(*a*) In the first step, fattyacyl CoA is converted to α, β-unsaturated fattyacyl CoA, by a dehydrogenase

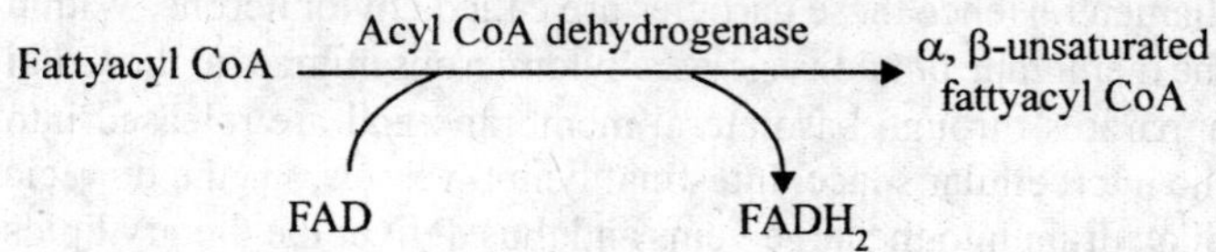

(*b*) Subsequently, α, β-unsaturated fattyacyl CoA is converted to β-hydroxyacyl CoA, with the addition of a molecule of water. This reaction is catalyzed by enoyl CoA hydratase.

(*c*) In the next step, β-hydroxyacyl CoA is converted to β-ketoacyl CoA, by NAD^+ dependent β-hydroxyacyl CoA dehydrogenase.

(*d*) In the last step, β-ketoacyl CoA is splitted to acetyl CoA and acyl CoA which has two carbon atoms less than the starting acyl CoA.

This reaction is catalyzed by thiolase and requires CoA. The process of β-oxidation is shown as follows :

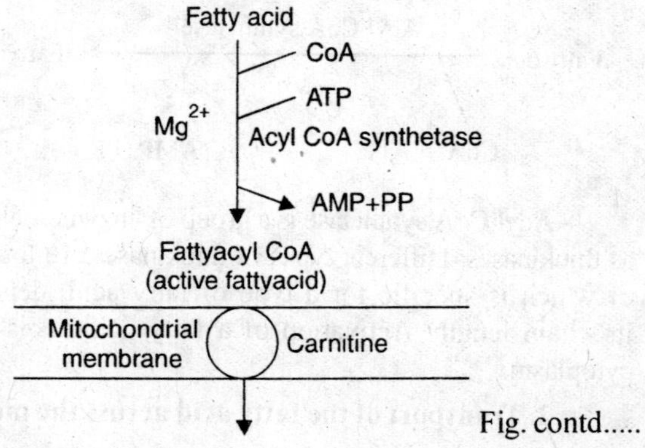

Fig. contd.....

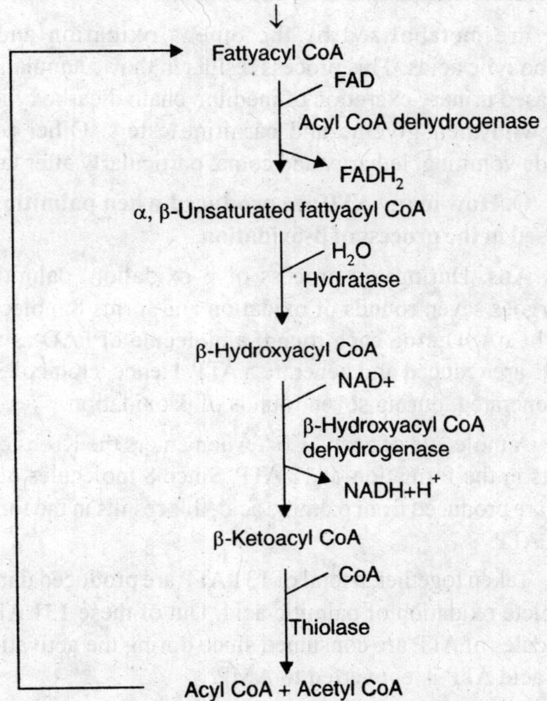

Q. Explain the role of carnitine in β-oxidation.

Ans. Fattyacids are oxidized in the mitochondria but the inner mitochondrial membrane is impermeable to fatty acyl CoA. Hence carnitine helps in the translocation of long chain fatty acids across the mitochondrial membrane.

Mitochondrial membrane has two enzymes referred to as carnitine palmitoyltransferase I and II (CPT-I and II). The outer mitochondrial membrane contains CPT-I which helps in the formation of acylcarnitine from fattyacyl CoA and carnitine. Acylcarnitine is translocated to the innerside of inner mitochondrial membrane and CPT-II helps in the formation of acyl CoA from acylcarnitine and CoA.

Q. What is sudden infant death syndrome (SIDS) ?

Ans. Sudden infant death syndrome (SIDS) is an inherited autosomal recessive disorder due to the deficiency of medium chain fattyacyl CoA dehydrogenase. This, in turn, results in the accumulation of medium chain fatty acids in various tissues where these are metabolized by the omega oxidation and form dicarboxylic acids. This process results in the accumulation and increased urinary excretion of medium chain dicarboxylic acids alongwith their glycine and carnitine esters. Other features include vomiting, lethargy and coma, particularly after fasting.

Q. How many ATP are produced when palmitic acid is oxidised in the process of β-oxidation.

Ans. During the process of β-oxidation, palmitic acid undergoes seven rounds of oxidation and forms 8 molecules of acetyl CoA. During each round, a molecule of FAD as well as NAD^+ are reduced and generate 5 ATP. Hence a total of 35 ATP are generated, during seven rounds of β-oxidation.

A molecule of acetyl CoA when enters the Krebs cycle, it results in the formation of 12 ATP. Since 8 molecules of acetyl CoA are produced from palmitic acid, this results in the formation of 96 ATP.

Taken together, a total of 131 ATP are produced during the complete oxidation of palmitic acid. Out of these 131 ATP, two molecules of ATP are consumed since during the activation of a fatty acid ATP is converted to AMP.

Hence, there is a net yield of 129 ATP, during the oxidation of a molecule of palmitic acid.

Q. What is α-oxidation ?

Ans. α-Oxidation is the process of oxidation of fatty acids where one carbon atom is removed from the carboxyl end of a fatty acid, as carbondioxide, sequentially. Neither acetyl CoA is formed nor ATP is generated during this process.

Q. What is Refsum's disease ?

Ans. α-Oxidation of some of the fatty acids, such as phytanic acid, requires a mono-oxygenase which is an α-hydroxylating enzyme. Inherited deficiency of this enzyme results in the accumulation of phytanic acid leading to a neurological condition, referred to as Refsum's disease.

Q. What is ω-oxidation ?

Ans. ω-Oxidation is the process of oxidation of fatty acids from both the ends, simultaneously. During this process the omega carbon atom, *i.e.* the carbon atom of the end methyl group, is oxidised, leading to the formation of a dicarboxylic acid. This reaction is catalyzed by ω-hydroxylase. The dicarboxylic acid, so formed can undergo β-oxidation from both the ends.

Q. How a fatty acid with odd number of carbon atoms is oxidised ?

Ans. Fatty acid containing odd number of carbon atoms is oxidised by the process of β-oxidation. This however, leads to the formation of propionyl CoA, instead of acetyl CoA, at the end of the reaction cycle. Propionyl CoA is subsequently converted to succinyl CoA via methylmalonyl CoA and is used in the Krebs cycle.

Q. Describe metabolic fates of acetyl CoA ?

Ans. Acetyl CoA mainly enters the Krebs cycle. Besides Krebs cycle, acetyl CoA is also used for the synthesis of ketone bodies, cholesterol and fatty acids.

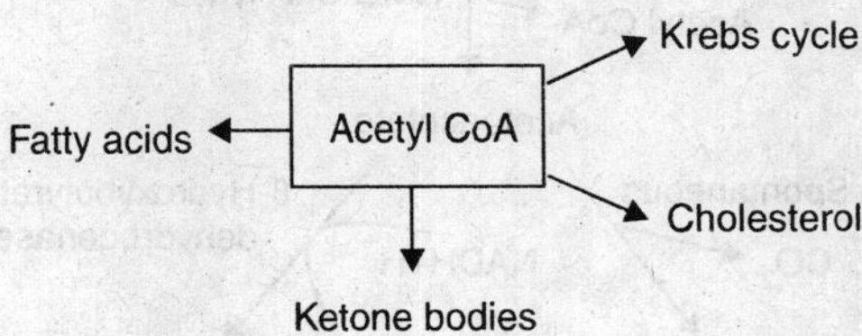

Q. What are ketone bodies ?

Ans. Acetoacetate, β-hydroxybutyrate and acetone are collectively referred to as ketone bodies. Out of these three, acetoacetate is first synthesised as a parent ketone body. This is subsequently decarboxylated to acetone or reduced to β-hydroxybutyrate. Since acetoacetate and acetone have ketone groups, these substances are referred to as ketone bodies. Ketone bodies can be used as a source of energy during starvation.

Q. Describe the process of synthesis of ketone bodies ?

Ans. Ketone bodies are synthesised from acetyl CoA in the liver, as follows :

1. Two molecules of acetyl CoA condense together and form acetoacetyl CoA. This reaction is catalysed by thiolase.

2. Acetoacetyl CoA further condenses with another molecule of acetyl CoA and forms β-hydroxy-β-methylglutaryl CoA (HMG CoA). This reaction is catalysed by HMG-CoA synthase.

3. In the next step, HMG-CoA lyase removes a molecule of acetyl CoA and converts HMG-CoA to acetoacetate.

4. Subsequently, acetoacetate in either spontaneously decarboxylated to acetone or reduced to β-hydroxybutyrate by the dehydrogenase.

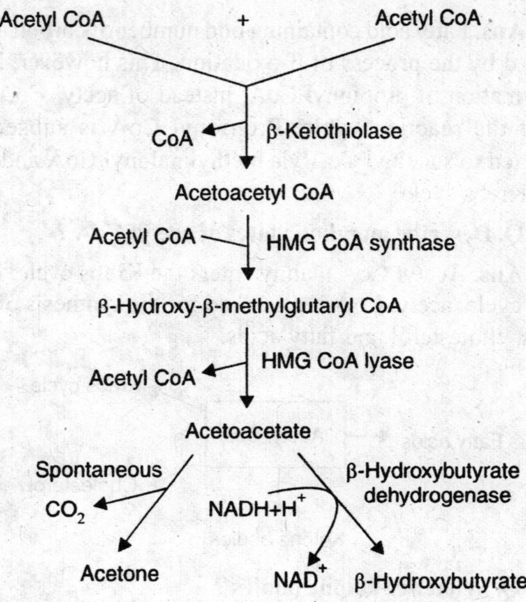

Q. Describe the process of the catabolism of ketone bodies.

Ans. 1. β-hydroxybutyrate is first oxidized to acetoacetate, by the dehydrogenase.

2. Subsequently, acetoacetate accepts a molecule of co-enzyme A, from succinyl CoA and is converted to acetoacetyl CoA. This reaction is catalysed by thiophorase.

This enzyme is not present in the liver but is found in the peripheral tissues.

3. Acetoacetyl CoA, thereafter, is degraded to two molecules of acetyl CoA. This reaction is catalysed by thiolase.

Q. What is ketosis ? Describe the conditions which lead to ketosis.

Ans. Increased production, accumulation and excretion of ketone bodies is referred to as ketosis. Increased accumulation of ketone bodies in the blood is referred to as ketonemia. Their increased excretion in urine is called ketonuria.

Normally, their blood concentration is 2-3 mg/dl. However, their production is increased during prolong starvation, sever-diabetes mellitus or on a high-fat, low-carbohydrate diet. This is due to increased oxidation of fatty acids which are used as a source of energy and lead to increased production of acetyl CoA. Ketone bodies are produced in the liver.

Q. Why liver cannot utilize ketone bodies ?

Ans. Due to the absence of thiophorase, liver can not utilise ketone bodies. This enzyme is present in the peripheral tissues such as brain, muscle, etc.

Thiophorase activates acetoacetate, with the transfer of CoA from succinyl CoA. Acetoacetyl CoA thereafter can be degraded to two molecules of acetyl CoA by the thiolase.

Q. What is diabetic ketoacidosis ?

Ans. Diabetic ketoacidosis occurs due to the severe deficiency of insulin with excess of glucagon and other hyperglycemic hormones. It is a common feature in patients with insulin-dependent diabetes mellitus. It is characterized by hyperglycemia, ketonemia, ketonuria as well as water and electolyte imbalance. This , in turn, can be corrected by insulin administration.

Q. Describe the process of the biosynthesis of phospholipids.

Ans. During biosynthesis of phospholipids such as phosphatidylethanolamine or phosphatidylcholine, the nitrogenous base is first activated by a kinase. Phosphoethanolamine or phosphocholine then reacts with CTP and forms CDP-ethanolamine or CDP-choline which subsequently, reacts with 1, 2 diacyl glycerol and forms phosphatidylethanolamine or phosphati-dylcholine.

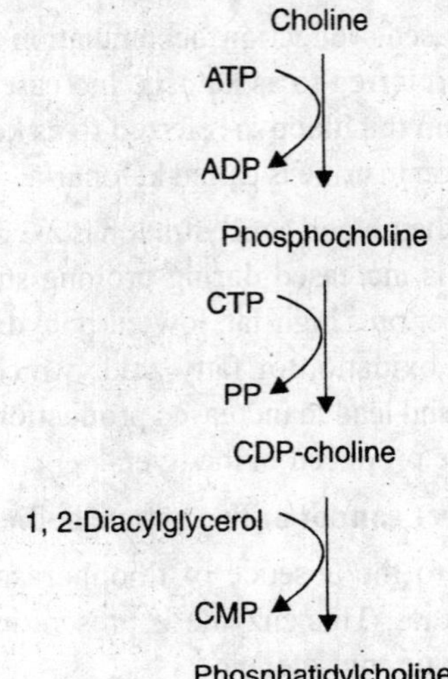

Q. What is physiological significance of lecithin ?

Ans. The ratio of lecithin to sphingomyelin (L : S ratio), in the amniotic fluid increases with gestation and is an index of foetal lung maturity. At 34 weeks of gestation, L : S ratio is > 2.

Q. What is sphingosine ?

Ans. Sphingosine is an aminoalcohol found in sphingolipids.

Q. What is Niemann-Pick disease ?

Ans. Niemann-Pick disease is due to deficiency of enzyme sphingomyelinase. It results in the excessive accumulation of sphingomyelins in the blood. There is enlargement of liver and spleen, with mental retardation.

Q. What is Tay-Sach's disease ?

Ans. Tay-Sach's disease is an autosomal recessive lipid storage disease due to the deficiency of hexosaminidase A. It results in the accumulation of gangliosides GM_2, in the neuronal cells. Infants become progressively weak with mental retardation and cherry red spot on macula, and usually die by the age of 3 years.

Q. What is Gaucher's disease ?

Ans. Gaucher's disease is an inherited disorder of lipid storage due to the deficiency of glucocerebrosidase. It results in the deposition of glucocerebrosides in the macrophages of the reticuloendothelial system. Common features of the disease include hepatomegaly, splenomegaly, thrombocytopenia, anemia, and erosions of long bones and pelvis.

Q. What is Krabbe's disease ?

Ans. Krabbe's disease is also called as globoid leucodystrophy. It is due to the deficiency of lysosomal galactocerebrosidase. It results in the accumulation of galactocerebrosides in the white matter, leading to absence of myelin, and mental retardation.

Q. What is Fabry's disease

Ans. Fabry's disease is due to the deficiency of α-galactosidase A. This results in the accumulation of ceramidetrihexosides in the kidney. This is characterized by pain in lower extremeties, skin rashes and renal failure.

Q. Describe sphingolipid storage diseases.

Ans. Sphingolipid storage diseases, also referred to as sphingolipidoses, is a group of disorders due to the genetic deficiency of some sphingolipid catabolizing enzyme resulting in the accumulation of a particular substrate in the lysosomes of the specific tissues. These are transmitted as recessive genetic abnormalities, *e.g.* Niemann-Pick disease, Tay-Sach's disease, etc.

Q. What are triacylglycerols ? How these are degraded ?

Ans. Triacylglycerols, also called triglycerides, are triacylesters of glycerol. These are mainly stored in the adipose tissue and provide a source of energy during starvation.

These are the principle sources of fat stored in the body.

The process of catabolism of triglycerides in the adipose tissue is referred to as lipolysis. Lipolysis is carried out by a group of enzymes called lipases. There are two main types of lipases referred to as lipoprotein lipase and hormone-sensitive lipase.

Lipoprotein lipase – It is located on the surface of the endothelial cells of capillaries and hydrolyse fatty acids from α-carbon atom of triacylglycerols, present in lipoproteins chylomicronds and VLDL. Lipoprotein lipase requires apo C-II and is activated by insulin.

Hormone sensitive lipase – It is specific for the hydrolysis of fatty acids from triglycerides that are stored in the adipose tissue. It is a regulatory enzyme. Its activity is stimulated by lipolytic hormones such as epinephrine, glucagon and ACTH. On the other hand, insulin and prostaglandins inhibit its activity.

Catabolism of triglycerides results in sequential hydrolysis of fatty acids from triglycerides and formation of glycerol and free fatty acids.

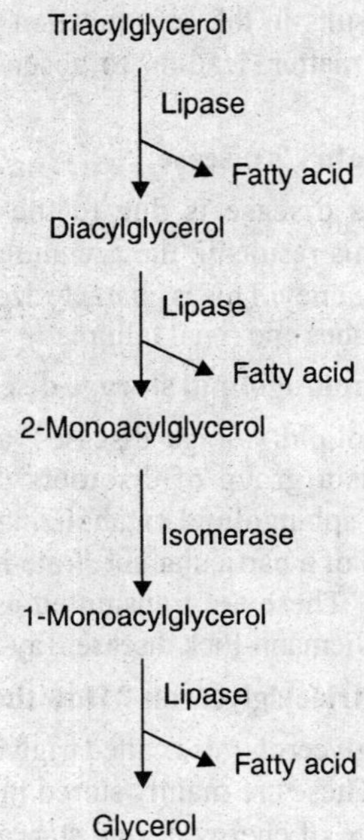

Q. Describe the process of the biosynthesis of triglycerides.

Ans. Biosynthesis of triacylglycerols is referred to as lipogenesis. It occurs in several tissues but mainly in the liver and adipose tissue.

Biosynthesis of triglycerides requires α-glycerolphosphate and the activation of fatty acids.

1. **Synthesis of glycerolphosphate.** Glycerolphosphate can be synthesized from glycerol by glycerol kinase which is present in the liver, kidney, and lactating mammary glands.

2. Fattyacyl CoA combines with glycerolphosphate which

is linked at 1-carbon atom of glycerol and forms 1-acylglycerol-3-phosphate. This reaction is catalyzed by acyltransferase.

3. Another molecule of acyl CoA is added to 1-acyl-glycerol-3-phosphate, by the same enzyme and forms 1, 2-diacyl-glycerolphosphate.

4. A molecule of phosphoric acid is removed from 1, 2-diacylglycerolphosphate and 1, 2-diacylglycerol is formed. This reaction is catalyzed by phosphatase.

5. A third molecule of acyl CoA then links to diacylglycerol and forms triacylglycerol.

Q. Explain the process of De novo synthesis of fatty acids ?

Ans. The process of synthesis of fatty acids from acetyl CoA is referred to as De novo synthesis. It occurs in the cytosol, as follows :

1. **Transport of acetyl CoA from the mitochondria to the cytosol** – Acetyl CoA is produced in the mitochondria where it condenses with oxaloacetate and forms citrate. This reaction is catalyzed by citrate synthase.

Citrate is freely permeable and thus comes out from the mitochondria to the cytosol. In the cytosol, citrate lyase cleaves citrate to oxaloacetate and acetyl CoA.

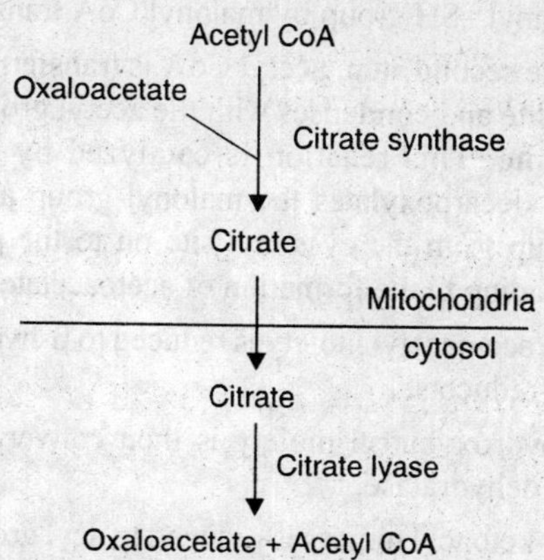

2. **Conversion of acetyl CoA to malonyl CoA** – Within the cytosol, acetyl CoA is converted to malonyl CoA. This reaction is catalyzed by acetyl CoA carboxylase and requires biotin.

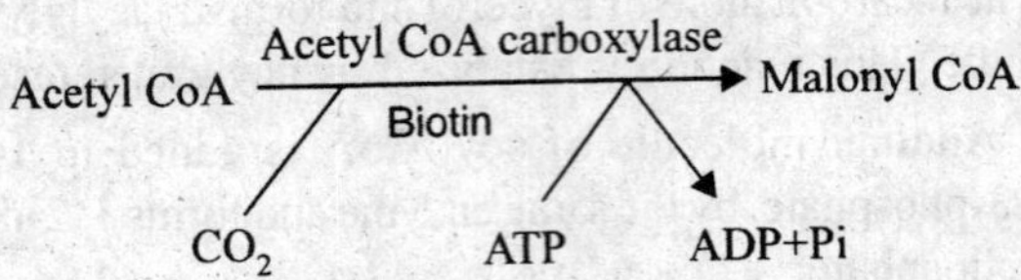

3. **Synthesis of palmitic acid** – Fatty acid synthesis requires an enzyme complex, referred to as fatty acid synthase.

In eukaryotes, fatty acid synthase consists of two subunits, each one of which has enzyme activities. These are referred to as acetyl CoA transacylase, malonyl CoA transacylase, condensing enzyme, β-ketoacyl reductase, β-hydroxyacyl dehydratase, enoyl reductase dehydratase, and palmitoyl thioesterase. It also has a domain that covalently binds a molecule of 4-phosphopantetheine (a derivative of pantothenic acid). In prokaryotes fatty acid synthase is a multienzyme complex where the 4-phosphopantotheine domain is a separate potein referred to as acyl carrier protein (ACP). Due to the presence of two molecules of ACP (one in each subunit), the enzyme has two–SH groups which are referred to the cysteinyl –SH group and the panto-theinyl –SH group.

Synthesis of palmitic acid occurs in five steps :

(*i*) In the first step, acetyl CoA is linked to the cysteinyl–SH group of the enzyme complex. This reaction is catalyzed by acetyl CoA transacylase. Thereafter, a molecule of malonyl CoA is linked to the pantotheinyl –SH group by malonyl CoA trans-acylase.

(*ii*) In the second step, acetyl CoA is transferred form the cysteinyl–SH site and condenses with the acetyl group on to the pantotheinyl site. This reaction is catalyzed by condensing enzyme which decarboxylates the malonyl group and transfers this acetyl group from the cysteinyl site on to the pantotheinyl site, thereby leading to the formation of acetoacetate.

(*iii*) The acetoacetyl moiety is reduced to β-hydroxybutryl, by β-ketoacyl reductase.

(*iv*) β-hydroxybutryl moiety is then converted to α, β-ketoacyl by a dehydratase.

(*v*) α, β-ketoacyl moiety is reduced to acyl group by enoyl reductase.

With the sequential additions of the two carbon units, in seven rounds, palmitate is released by the action of palmitoyl thioesterase.

Q. How De novo synthesis of fatty acid differs from fatty acid synthesis in the mitochondria and the endoplasmic reticulum ?

Ans. De novo synthesis of fatty acid occurs in the cytosol. In this process, acetyl CoA is used. On the other hand, chain elongation of the pre-existing fatty acids takes place in the mitochondria as well as endoplasmic reticulum,.

In the mitochondria, chain elongation occurs with the successive additions of the acetyl CoA subunits, by the process which is the reversal of β-oxidation. In this process, both NADH as well as NADPH are used.

In the endoplasmic reticulum, chain elongation occurs by the process similar to that exists in the cytosol. In this process malonyl CoA is used as a source of two carbon units.

Differences in the three processes are as follows :

Cytosol	Mitochondria	Endoplasmic reticulum
1. There is actual synthesis of fatty acids from acetyl CoA	Chain elongation of palmitate to higher chain fatty acids takes place, by utilizing acetyl CoA	Chain elongation takes place with the addition of malonyl CoA
2. Synthesis requires multienzyme complex which is referred to as fatty acid synthase. It has two sub-units, each one of which comprises of seven enzymes and acyl carrier protein (ACP)	Fatty acid synthase complex is not required	Enzyme complex is not required
3. NADPH is used in the reduction process	Both, NADH as well as NADPH are used in two different steps	NADPH is used

Q. How fatty acid synthesis is regulated in the cytoplasm ?

Ans. De novo synthesis of fatty acids is regulated as follows :

1. **By acetyl CoA carboxylase** – Conversion of acetyl CoA to malonyl CoA is a rate limiting step. Acetyl CoA carboxylase is allosterically regulated by citrate, which activates the enzyme. It is also stimulated by isocitrate and α-ketoglutarate. Insulin stimulates its activity while glucagon inhibits it.

2. **By fatty acid synthase** – High carbohydrate diet as well as fat free diet stimulate the synthesis of fatty acid synthase. On the other hand, high fat diet and starvation reduce enzyme synthesis.

Glucose-6-phosphate and other phosphorylated sugars also cause allosteric activation of fatty acid synthase.

Q. Describe various metabolic fates of cholesterol ?

Ans. Cholesterol is a precursor of :

(*i*) Steroid hormones,

(*ii*) Vitamin D, and

(*iii*) Bile acids.

Q. How is cholesterol synthesized in the body ?

Ans. Cholesterol is synthesized from acetyl CoA, as follows:

1. In the first step, two molecules of acetyl CoA combine together and form acetoacetyl CoA. This reaction is catalyzed by thiolase.

2. Acetoacetyl CoA further combines with another molecule of acetyl CoA and forms β-hydroxy-β-methylglutaryl CoA (HMG CoA). This reaction is catalyzed by HMG CoA synthase.

3. HMG CoA is reduced to mevalonic acid. This reaction is catalyzed by NADPH dependent HMG CoA reductase.

It is a regulatory enzyme.

4. Mevalonic acid is decarboxylated and phosphorylated to an isoprenyl unit, referred to as isopentenylpyrophosphate.

There occurs stepwise addition to two phosphate molecules (from ATP) by kinase and decarboxylation by decarboxylase.

5. The isoprenyl unit is isomerized, by an isomerase. In this reaction a molecule of isopentenylpyrophosphate is

isomerized to dimethyl allylpyrophosphate.

Subsequently, these two isoprenoid units (isopentenyl-pyrophosphate and dimethylallylpyrophosphate) condense together and form geranylpyrophosphate.

6. In the next step, another molecule of isopentenyl pyrophosphate condenses with geranylpyrophosphate and forms farnesylpyrophosphate. This reaction is catalyzed by geranyl transferase.

7. Thereafter, two molecules of farnesylpyrophosphate condense and form squalene.

Squalene has 30 carbon atoms. This reaction is catalyzed by squalene synthase. Subsequently, there is ring closure and three carbon atoms are removed as methyl groups. Squalene is thus converted to cholesterol which has twentyseven carbon atoms.

Q. How chotesterol biosynthesis is regulated ?

Ans. Cholesterol biosynthesis is regulated by various processes :

1. **By feed back inhibition.** Cholesterol inhibits the activity of HMG CoA reductase, by reducing the synthesis of the enzyme as well as its inactivation.

2. **By dietary cholesterol.** Amount of dietary cholesterol affects the rate of De novo synthesis of cholesterol. High intake of cholesterol reduces its synthesis.

3. **By the removal of cholesterol from the liver.** Liver removes cholesterol by three different processes :

(*i*) It helps in the esterification of cholesterol, by lecithin-cholesterol acyltransferase (LCAT). Esterified cholesterol is incorporated into HDL.

(*ii*) In the liver, cholesterol is converted to bile acids. Both bile acids as well as free cholesterol are excreted in the bile.

(*iii*) In the liver, cholesterol is also incorporated into VLDL and HDL.

Q. What is atherosclerosis ?

Ans. Atherosclerosis is a disease of arterial wall due to the deposition of cholesterol within tunica intima. Risk of atherosclerosis is directly related to the plasma concentration of

LDL-cholesterol (which is also referred to as bad cholesterol). High concentration of HDL-cholesterol (which is also referred to an good cholesterol) prevents atherosclerosis. High calorie diet, hypercholesterolemia, obesity, hypertension and smoking are the risk factors for atherosclerosis.

Q. What are lipoproteins ? How are they classified ?

Ans. Lipoproteins are hydrophilic complexes of lipids with proteins.

Lipoproteins are classified into four groups, depending upon their density and lipid as well as protein contents, as follows :

1. **Chylomicrons** – They have the lowest density, very high content of fat (mainly triglycerides) with small amount of proteins. Main apoproteins present in chylomicrons include B-48, A-I and A-II. These are the transporters of dietary fat.

2. **Very low density lipoproteins (VLDL)** – VLDL particles have lipids, mainly triglycerides, which are of endogenous origin. VLDL contains Apo B-100. Its density is more than chylomicrons but lower than LDL.

3. **Low density lipoproteins** – LDL is synthesized from VLDL. It is rich in cholesterol and transports free cholesterol from the liver to the peripheral tissues.

4. **High density lipoproteins (HDL)** – HDL carries cholesterol from peripheral tissues to the liver for its utilization there. It also contains LCAT enzyme which helps in the esterification of cholesterol.

Q. Describe the metabolism of various lipoproteins.

Ans. Metabolism of lipoproteins takes places, as follows :

Metabolism of chylomicrons

Chylomicrons are released from the mucosal cells of the intestine as nascent chylomicrons. They carry apo B-48 and acquire apo C and apo E from HDL. As a result of the activation of lipoprotein lipase by apo C-II, triacylglycerols are hydrolyzed. Apo C is returned back to HDL and chylomicrons, as chylomicron remnants, bind to the receptors on the hepatocytes and are internalized therein.

Metabolism of VLDL

VLDL particles are released as nascent VLDL from the liver and carry endogenous triglycerides, apo B-100 and apo A-I. They acquire apo C-II and apo E from the circulating HDL and form mature VLDL. As a result of the activation of lipoprotein lipase, triglycerides are hydrolysed. Free fatty acids are released to the circulation whereas apo C and apo E are transferred to HDL. VLDL also transfers its phospholipids to HDL and receive cholesterol from there. Due to the exchange of these lipids, their particle size is reduced while density is increased and VLDL are converted to VLDL remnants, also referred to as IDL. They acquire more cholesterol and form LDL.

Metabolism of LDL

LDL binds to receptors and are internalized by endocytosis. They carry cholesterol, high concentration of which is atherogeric.

Metabolism of HDL

HDL helps in the metabolism of chylomicrons and VLDL. It contains apo C-II, an activator of lipoprotein lipase and LCAT which esterifies cholesterol. HDL thus removes excess of cholesterol from the peripheral tissues and transports it to the liver for its utilization there.

Q. Describe hyperlipoproteinemias.

Ans. Hyperlipoproteinemia is a group of disorders due to increased concentration of various lipoproteins. Depending upon the rise in the type of lipoprotein, they may be classified into different groups, as follows—

Type I hyperlipoproteinemia – This is due to an increase in chylomicrons and also referred to as exogenous hyperlipidemia. Plasma triglyceride concentration is increased which may be a result of the deficiency of either lipoprotein lipase or apo C-II.

Type II hyperlipoproteinemia – It may result in increased levels of either LDL or both LDL as well as VLDL. This, in turn, increases cholesterol either alone (hypercholesterolemia) or with triglycerides (combined hyperlipidemia).

Type III – It is also called as dysbetalipoproteinemia. Plasma triglyceride level is increased due to the increased concentration of LDL.

Type IV – It is due to increase in VLDL and is also referred to as hyperprebetalipoproteinemia.

Type V – There is a rise in triglycerides which is due to increased levels of both chylomicrons as well as VLDL. It is also referred to as mixed hyperlipidemia.

Q. What is tangier's disease ?

Ans. Tangier's disease is due to the deficiency of apo A which, in turn, leads to the deficiency of HDL.

Q. What is abetalipoproteinemia ?

Ans. Abetalipoproteinemia is due to the complete deficiency of apo B which, in turn, results in a reduction in VLDL and LDL. Plasma lipids, particularly, triglyceride levels are reduced.

6
Proteins

Q. What are proteins ?

Ans. Protein is the class of natural products containing nitrogen. These are essential for life and form an important structural component of the cell. Proteins are composed of small units called amino acids which are linked together by peptide bonds.

Q. Differentiate between precipitation, denaturation and coagulation of proteins.

Ans. Precipitation : Due to their colloidal nature in solution, protein can be precipitated by dehydration or neutralization of the polar groups.

Proteins can be precipitated when pH of the solution is brought to iso-electric pH. It is also referred to as iso-electric point (pI). Protein can also be precipitated by neutral salt such as ammonium sulphate, by the process referred to as salting out. Certain organic solvents, such as alcohol, are also used as precipitating agents.

Denaturation : Precipitation by certain physical agents such as heat and radiation or certain chemical agents such as heavy metals, alkaloidal reagents, etc. results in the loss of biological activity of a protein. This, in turn, occurs due to the loss of secondary, tertiary and quaternary structures, thereby causing denaturation. Heavy metals (*e.g.* lead and mercury), alkaloidal reagents (*e.g.* picric acid), strong acids (*e.g.* trichloroacetic acid) or strong alkalies (*e.g.* NaOH) are often used for the preparation of protein-free filtrate of blood.

Coagulation of proteins : Irreversible denaturation, particularly by heat at iso-electric pH, results in the formation of semisolid viscous precipitate of the protein referred to as coagulum, which floats on the surface.

Q. What is electrophoresis ?

Ans. Electrophoresis is a technique for the separation of the charged particles by means of electrical current. A charged particle, when placed in the electric field migrates, either towards anode or cathode. Depending upon the charge, particle size or shape, different molecules migrate at different speeds. Movement of the charged particles takes place in a buffered solution, on some inert material such as filter paper, cellulose strip or agar gel.

Q. Describe the pattern of serum proteins in a normal individual.

Ans. Different fractions of serum proteins can be separated by paper electrophoresis by using 0.1 molar barbitone buffer (pH 8.6). At this pH, all the proteins carry a net negative charge and migrate towards anode. When a constant current is applied (1 - 2 mA per strip), different proteins migrate at different rates. Their positions can be marked as separate bands, after staining with a suitable dye such as bromophenol blue.

Pattern of serum proteins helps in the diagnosis of a disease.

In a normal serum, five bands are observed which represent albumin, α_1–globulin, α_2–globulin, β–globulin and γ–globulin. In patients with nephrotic syndrome, albumin band is reduced whereas in a case of multiple myeloma a dense band between a– and β–globulins is observed. In patients with chronic infections, γ-globulin band is more prominent.

Q. Describe the process of digestion of protein.

Ans. Dietary proteins are hydrolyzed by a number of enzymes in the gastrointestinal tract and small intestine. Digestion starts in the stomach by the action of pepsin. Pepsin is secreted as pepsinogen by the chief cells of the mucosa. Pepsinogen is activated by H^+ ions (HCl) of the gastric mucosa and subsequently by autocatalysis.

Pepsin catalyzes the hydrolysis of the peptide linkage where amino group is contributed either by an aromatic or acidic amino acid.

Hydrolyzed products of pepsin pass-on to small intestine and are hydrolyzed further, by several proteases which are secreted by the pancreas. These include trypsin, chymotrypsin, carboxypeptidases, aminopeptidases and dipeptidases.

Trypsin is secreted as trypsinogen. Trypsinogen is converted to trypsin by an enteropeptidase, also referred to as enterokinase. Trypsin hydrolyses peptide linkages which are contributed by the carboxylate group of basic amino acids.

Chymotrypsin is secreted as chymotrypsinogen which is activated by trypsin. Chymotrypsin hydrolyzes peptide bonds where carboxyl group is contributed by the aromatic amino acids, leucine, methionine, asparagine or histidine.

These enzymes are referred to as endopeptidases since they attack peptide linkages, endogenously (within the peptide chain). Besides, carboxypeptidases and aminopeptidases remove C-terminal and N-terminal amino acids exogenously. These are called as exopeptidases. Dipeptides are hydrolyzed by dipeptidases. The end product of the complete digestion of a protein is thus a mixture of amino acids.

Q. Describe some biologically active peptides.

Ans. Some of the biologically active polypeptides are glutathione, oxytocin, vasopressin and insulin.

Glutathione : Glutathione (GSH) is a tripeptide consisting of glycine, cysteine and glutamic acid. It is a reducing agent and is important in the detoxification of H_2O_2, particularly in the erythrocytes. Reduced glutathione gets oxidised. It is regenerated with NADPH + H^+, which is produced in the HMP shunt.

Oxytocin : Oxytocin is a hormone secreted by the posterior pituitary. It is a nanopeptide. Oxytocin acts on the uterine muscle before parturition. It also helps in the ejection of milk.

Vasopressin : Vasopressin, also called as antidiuretic hormone (ADH), is a nanopeptide. It is secreted by the posterior pituitary and helps in the reabsorption of water in the collecting duct.

Insulin : It is a polypeptide consisting of 51 amino acids. It has two peptide chains, *i.e.* an α–chain with 30 amino acids and a β–chain with 21 amino acids. Two chains are linked together by two disulphide linkages. It is secreted by the β-cells of the islets of Langerhans. Insulin helps in lowering blood glucose level.

Q. Describe structural organization of protein.

Ans. Structural organization of a protein refers to the order of different amino acids and the manner in which these are linked. There are four levels of protein structure —

1. **Primary structure of protein :** Primary level of protein structure refers to the order and sequence of covalently linked amino acid residues in a polypeptide chain. Primary structure of a protein has peptide bonds and disulphide linkages.

2. **Secondary level of protein structure :** Secondary level of protein structure refers to folding and twisting patterns of the polypeptide chain. Various forces which stablize secondary level of protein structure include hydrogen bonds, hydrophobic and electrostatic interactions, and Van-der-Waals forces. Secondary structure include α-helix and β-pleated sheet structure.

3. **Tertiary level of protein structure :** Tertiary level of protein structure refers to intramolecular folding of the polypeptide chain to a compact, three-dimensional structure with a specific shape. Hydrophobic interactions are major forces in the maintenance of tertiary structure of protein.

4. **Quarternary structure of protein :** Quarternary level of protein structure refers to spatial arrangment of several similar or dissimilar peptides. These proteins have more than one subunit and are referred to as oligomeric proteins, *e.g.* hemoglobin, a tetramer with 2α and 2β subunits.

Q. Describe α-helical structure of a protein.

Ans. Folding and twisting of the backbone of a polypeptide chain results in the formation of the coiled or helical structure of the protein molecule. As a result of intramolecular folding, hydrogen bonds are formed.

The polypeptide helix may either be right-handed, (referred to as α-helix) or left-handed, (referred to as β-helix). Most of the naturally occuring proteins are right-handed and possess α-helical structure. It is a more stable conformation and is formed spontaneously. Each turn of the α-helix has 3.6 amino acid residues, *e.g.* α-keratin.

Q. Describe β-pleated sheet-like structure of proteins?

Ans. The β-structure of a protein refers to the conformation where a protein may contain several polypeptides which are linked together through hydrogen bonds and form a pleated sheet-like structure. The β-sheet orientation is formed due to stretching of the helix of the polypeptide chain. It results in the formation of hydrogen bonds between neighbouring polypeptides rather than within one polypeptide chain.

This is referred to as β-structure since it was described as the second regular structure after the α-helix, *e.g.* wool.

Q. Differentiate between globular and fibrous proteins?

Ans. Tertiary structure of a protein may result in compact three-dimensional morphology giving it a characteristic shape. A protein having the spheroidal shape is referred to as globular protein. These have relatively high water solubility with a variety of functions, *e.g.* hemoglobin.

On the other hand, fibrous proteins exhibit a long cylindrical shape (rod-like or fibre shape). These have relatively low solubility in water with a characteristic structural role in the cell, *e.g.* collagen.

Q. Classify proteins according to their solubility.

Ans. Depending upon their solubility and other physical properties, proteins can be classified into three groups:

1. Simple proteins,
2. Conjugated proteins, and
3. Derived proteins

SIMPLE PROTEINS

These are polypeptides which are soluble in water, acid or alkali. Depending upon their solubility, simple proteins are further subdivided into several groups, as —

1. **Albumins :** Albumins are soluble in water as well as a salt solution, *e.g.* serum albumin, lactalbumin, egg albumin, etc.

2. **Globulins :** These are sparingly soluble in water but clearly soluble in a neutral salt solution, *e.g.* globulin present in milk, serum or egg.

3. **Glutelins :** These are soluble in dilute acid or alkali, *e.g.* glutenin found in wheat.

4. **Prolamines :** These are insoluble in water but soluble in 6 to 20% alcohol, *e.g.* gliadin of wheat or zein of maize.

5. **Scleroproteins :** These are highly insoluble in water but

partially soluble in hot and strong alkaline solution. These are mainly present in a supporting tissue such as bone and cartilage or in a protective tissue like hair, nail, etc. For example, keratin of nail.

6. **Histones :** These are soluble in water and rich in basic amino acids. These are generally found within the nucleus, in association with DNA. *e.g.* nucleohistones.

CONJUGATED PROTEINS

These are simple protein but associated with some non-protein component.

For example, **glycoproteins :** Proteins conjugated with carbohydrates.

Nucleoproteins : Proteins conjugated with nucleic acids

Metalloproteins : Proteins conjugated with a metal, *e.g.* hemoglobin.

DERIVED PROTEINS

These are the derivatives of simple proteins and may include —

1. Primary derived proteins : These are denatured or coagulated proteins, and

2. **Secondary derived proteins :** These are produced due to cleavage of peptide bonds, *e.g.* peptones and proteoses.

Q. Classify proteins according to their biological functions.

Ans. According to their biological functions proteins are classified into different groups, as follows —

1. **Structural proteins :** These proteins form an essential part of a particular structure in the body, *e.g.* collagen.

2. **Enzymes :** These are the proteins which catalyze biochemical reactions in a living organism *e.g. pepsin.*

3. **Hormones :** Some of the proteins regulate metabolic processes of the body and are referred to as hormones, *e.g.* insulin.

4. **Transport proteins :** Several proteins serve as carriers for the transport of different substances, *e.g.* hemoglobin, albumin, etc.

5. **Receptors :** Several proteins act as receptors for the transport of substances across the cell membrane, *e.g.* LDL receptors.

6. **Antibodies :** Some of the proteins are produced in response to an antigen and provide immunity to the body. These are referred to as antibodies or immunoglobulins, *e.g.* γ – globulins.

7. **Storage proteins :** Some of the proteins bind with certain small molecular weight substances for their storage within the tissues, *e.g.* ferritin, ceruloplasmin, etc.

7

Amino Acids

Q. What is an amino acid ?

Ans. Amino acids are the hydrolyzed products of protein. Commonly found amino acids are α-amino acids. They have a primary amino group ($-NH_2$) and a carboxylic group (–COOH). Both the groups are present on the α–carbon atom.

Q. How amino acids are classified ?

Ans. Amino acids can be classified in several ways –

(*i*) According to the type of reaction of the R group,

(*ii*) According to the polarity of the side chain, and

(*iii*) According to nutritional requirement.

According to the type of reaction of the R group, amino acids are classified as –

(*a*) neutral amino acids (glycine, alanine, etc.), with an aliphatic or aromatic side chain.

(*b*) acidic amino acids (glutamic and aspartic acid), and

(*c*) basic amino acids (lysine, arginine, etc).

Q. What are the branched chain amino acids ?

Ans. These are aliphatic neutral amino acids which have a branch in their aliphatic side chain. These are valine, leucine, and isoleucine.

Q. What are hydroxyamino acids ?

Ans. Hydroxyamino acids are those amino acids which have a hydroxyl group (–OH) in their aliphatic side chain. These include serine and threonine.

These amino acids are important for the linkage of a protein with a prosthetic group, such as phosphoric acid, found in conjugaed proteins like casein.

Q. What are sulphur containing amino acids ?

Ans. In addition to C, H, O and N which are commonly found in amino acids, sulphur containing amino acids also contain sulphur, as a part of their aliphatic side chain, *e.g.* cysteine and methionine.

Q. What are aromatic amino acids ?

Ans. Aromatic amino acids have an aromatic ring. These include phenylalanine, tyrosine and tryptophan.

Q. What are neutral amino acids ?

Ans. Neutral amino acids are those amino acids which have one amino group (–NH_2) and one carboxyl group (–COOH). They exhibit a neutral reaction, *e.g.* alanine, leucine, etc.

These amino acids are also called monoamino - monocarboxylic acids.

Q. What are acidic amino acids ?

Ans. Amino acids which have one amino group but more than one (generally 2) carboxylic groups are called acidic amino acids, *e.g.* aspartic acid and glutamic acid.

These amino acids exhibit acidic reaction and are called monoamino-dicarboxylic acids.

Q. What are basic amino acids ?

Ans. Basic amino acids are those amino acids which have one carboxylic group with either more than one amino group in the aliphatic side chain or a heterocyclic ring which is basic in nature, *e.g.* lysine, arginine and histidine.

Q. What is an imino acid ?

Ans. Some of the amino acids have a secondary imino group (–NH–) instead of a primary α-amino group (–NH_2), in a heterocyclic ring, such as proline.

```
H2C ——— CH2
 |        |
H2C      CH – COOH            α
   \    /                 R – CH – COOH
     N                         |
     H                        NH2
```

Proline (Imino acid)   α-Amino acid

Q. How are amino acids absorbed ?

Ans. Absorption of amino acids is sodium dependent active process which requires specific transport proteins and

ATP. Glutathione and vitamin B_6 are also required in the absorption and transport of amino acids.

Several brush-border specific transport proteins have been found for the uptake of various amino acids –

1. For neutral amino acids with a short or polar side chain, *e.g.* for alanine, serine and threonine.

2. For neutral amino acids with aromatic or hydrophobic side chain, *e.g.*, phenylalanine tyrosine and methionine as well as branched chain amino acids.

3. For basic amino acids, *i.e.* lysine, arginine and histidine.

4. For acidic amino acids, *i.e.* aspartate and glutamate.

5. For imino acid, *e.g.* proline.

6. For β-amino acids, *e.g.* β-alanine. These amino acids normally are not found in a protein.

7. For dipeptides.

Though a protein is hydrolyzed to individual amino acids which are absorbed differently there also occurs a carrier for the transport of dipeptides which may be hydrolyzed within the intenstinal cell by dipeptidases.

Q. Describe some amino acid transport disorder ?

Ans. Hartnup's disease is a transport disorder where transport and absorption of neutral amino acids including tryptophan is reduced. It is a genetic defect of the transport protein. This may result in amino acid malabsorption and amino aciduria. Tryptophan deficiency, in turn, may also result in pellagra like features.

Q. Describe the processes of the removal of ammonia from amino acids.

Ans. The α-amino group from an amino acid can be removed as ammonia either by transamination or deamination.

Transamination

Transamination is a process of the transfer of an α-amino group from one amino acid to a keto acid resulting in the formation of a new amino acid and a keto acid. Such reactions are catalyzed by transaminases (aminotransferases). All amino acids except threonine and lysine undergo transamination.

During the process of transamination mainly α-ketoglutarate participate as an acceptor of the amino group.

Two most common transaminases found in the liver, muscle and several other tissues have diagnostic significance. These are referred to as GOT (glutamate oxaloacetate transaminase) and GPT (glutamate pyruvate transaminase). They catalyze a reversible reaction where either glutamate is used with oxaloacetate/pyruvate or glutamate is formed from alanine/aspartate with their reaction with α-ketoglutarate. GOT is also referred to as AST (aspartate aminotransferase) while GTP is also referred to as ALT (alanine aminotransferase).

Serum levels of these two enzymes, *i.e.* SGOT and SGPT are raised in cardiac and liver disorders. SGOT is more raised compared to SGPT, in a cardiac disease such as myocardial infarction while SGPT levels are more raised than SGOT in a liver disease such as hepatitis.

Deamination

Actual removal of the α-amino group, as ammonia, is referred to as deamination. Deamination may take place either with the removal of electrons and oxygen or without the transfer of electrons. These are referred to as oxidative deamination and nonoxidative deamination, respectively.

Oxidative deamination : Oxidative deamination involves transfer of electrons to oxygen and some acceptor such as FAD or NAD^+, etc. The process of oxidative deamination is catalyzed by several enzymes, *e.g.* L-amino acid oxidases. These enzymes catalyze oxidation of L-amino acids and contain FMN as a prosthetic group.

Non-oxidative deamination : Deamination of amino acids is catalyzed non-oxidatively without the transfer of electrons, *e.g.* serine, cysteine, histidine, etc. These are deaminated by serine dehydratase, desulfhydrase and histidase, respectively.

Q. What is transdeamination ?

Ans. Transdeamination is the process of the transamination followed by oxidative deamination. As α-amino group from most of the amino acids is transferred to α-ketoglutarate and results in the production of L-glutamate, which is finally removed by glutamate dehydrogenase, as ammonia.

Glutamate dehydrogenase catalyzes reversible set of

reactions and results either in the deamination of glutamate to ammonia and α-ketoglutarate or formation of glutamate from ammonia and α-ketoglutarate.

Q. Describe various processes of the removal of ammonia from the body.

Ans. Ammonia is synthesized in all the tissues. It is a highly toxic substance and is water insoluble. It is removed from the body from different tissues, as follows –

1. In the liver, ammonia is converted to urea by an enzymatic process of urea cycle which is the main excretory waste product of protein catabolism.

2. In the brain, ammonia is converted to glutamine by glutamine synthetase. Glutamine is transported through blood to the kidney and is converted to ammonia and glutamic acid by glutaminase. Glutamate is retransported while ammonia is excreted as NH_4^+.

Q. Outline urea cycle. What is the need for urea synthesis ? Where urea is synthesized ?

Ans. Urea cycle is a cyclic process, by which ammonia, a highly toxic water insoluble metabolite is converted to urea which is less toxic and water soluble excretory waste product. Urea is synthesized in the liver, as follows :

1. Ammonia reacts with CO_2 and forms carbamoylphosphate. This reaction is catalyzed by carbamoylphosphate synthetase. This enzyme requires ATP and N-acetylglutamate as a coenzyme.

2. Carbamoylphosphate combines with ornithine and forms citrulline via ornithine transcarbamoylase.

3. Citrulline combines with aspartate and forms argininosucinate. This reaction is catalyzed by argininosuccinate synthetase and requires ATP.

4. Argininosuccinate is degraded to fumarate and arginine. This reaction is catalyzed by arigininosuccinase.

5. Arginine is hydrolysed to urea and ornithine. This reaction is catalyzed by arginase. Urea is excreted through urine while ornithine re-enters the cycle.

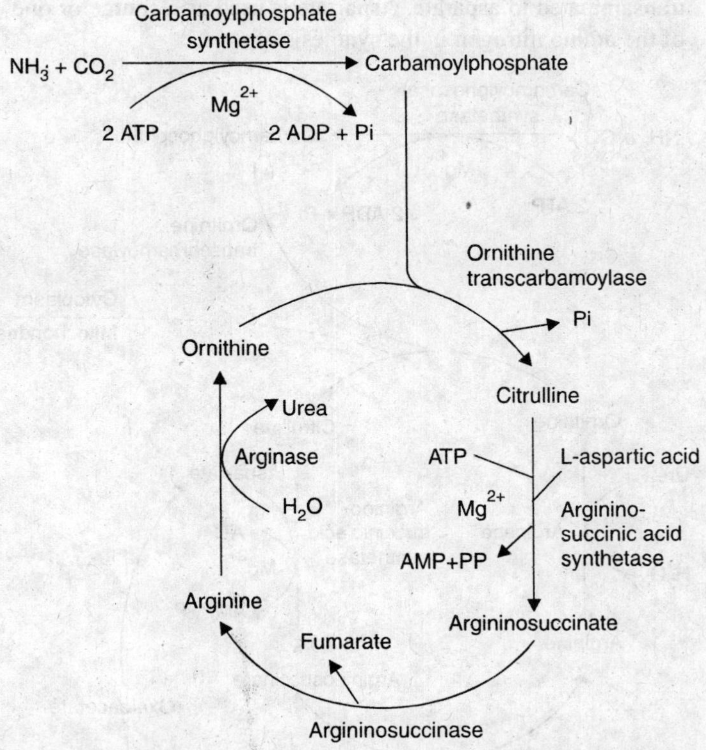

Q. What are the sources of nitrogens and carbon in urea ?

Ans. A molecule of urea has one carbon which is derived from Tactive carbondioxide (HCO_3^-). It also has two nitrogens one of which is derived from NH_3 while the other from aspartate.

Q. How urea synthesis is regulated ?

Ans. The first enzyme (carbamoylphosphate synthetase) of the urea cycle regulates the process of urea synthesis. It is allosterically activated by N-acetylglutamate. Other enzymes of the urea cycle are induced during protein catabolism.

Q. How is urea cycle linked to Krebs cycle ?

Ans. Urea cycle is also known as Krebs-Hanslet cycle. It is linked to Krebs cycle as fumarate generated in the urea cycle is used in the Krebs cycle and is converted to malate. Malate comes out of the mitochondria, changes to oxaloacetate and get transaminated to asparate. Aspartate is used as a source of one of the amino nitrogen in the synthesis of urea.

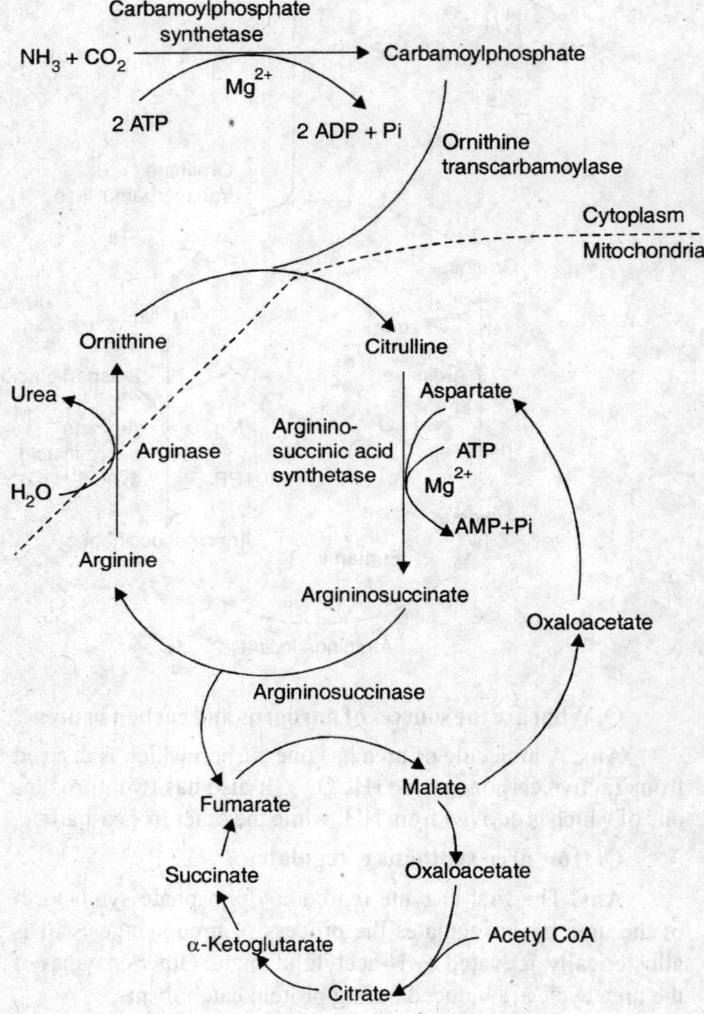

Some of the enzymes of urea cycle are located in the cytoplasm whereas argininosuccinic acid synthetase and arginino-succinase are present in the mitochondria. Hence citrulline,which is formed in the cytoplasm enters the mitochondria whereas arginine comes out of the mitochondria for its further entry in the cycle. A part of the cycle is thus operated in the cytoplasm whereas other in the mitochondria.

Q. Describe various metabolic disorders of the urea cycle?

Ans. Metabolic disorders of the urea cycle are:

1. Hyperammonemia type-I. It is due to increased level of ammonia in the blood, due to the lack of carbamoylphosphate synthetase. Ammonia being a highly toxic substance leads to mental retardation, blurring of vision and slurring of speech.

2. Hyperammonemia type-II. Increased levels of ammonia in blood may also occur due to the deficiency of ornithine trans-carbamoylase. Due to lack of this enzyme, carbamoylphosphate can not be used for the synthesis of citrulline. This in turn diffuses into the blood.

3. Citrullinemia. It is due to the deficiency of argininosuccinic acid synthetase. As a result of which citrulline levels are increased in the blood.

4. Argininosuccinic aciduria. Lack of argininosuccinase results in increased excretion of argininosuccinic acid in the urine.

5. Hyperargininemia. Due to lack of arginase, arginine levels are increased in the blood.

Q. What are metabolic fates of carbon skeletons of amino acids ?

Ans. After removal of the α-amino group from amino acids, their keto acids are either used in the synthesis of glucose or ketone bodies.

Catabolic products of some of the amino acids such as alanine, glutamate, aspartate, valine, etc., form pyruvate, α-ketoglutarate, oxaloacetate or succinyl CoA which enter the Krebs cycle and act as precursors of glucose. Therefore, these amino acids are referred to as glucogenic amino acids.

On the other hand, catabolism of leucine and lysine results in the formation of acetoacetyl CoA, a precursor of ketone bodies. Hence these amino acids are referred to as ketogenic amino acids.

The end products of the catabolism of some of the amino acids such as phenyalanine, tyrosine, tryptophan, isoleucine, etc., lead to the formation of both, the intermediates of Kreb cycle as well as precursor of ketone bodies, hence these amino acids are referred as glucogenic as well as ketogenic.

Q. Describe metabolic fates of glycine.

Ans. Glycine is a precursor of several important compounds :

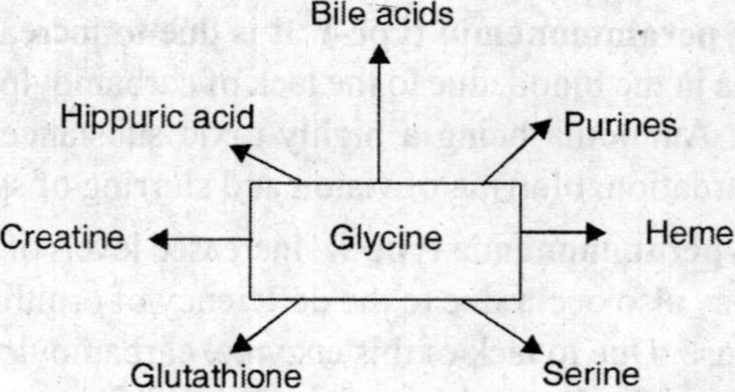

1. Serine is formed from glycine in a reversible reaction which is catalyzed by serine hydroxymethyltransferase. N^5, N^{10} -methylene. FH_4 is used as a source of one-carbon moiety.

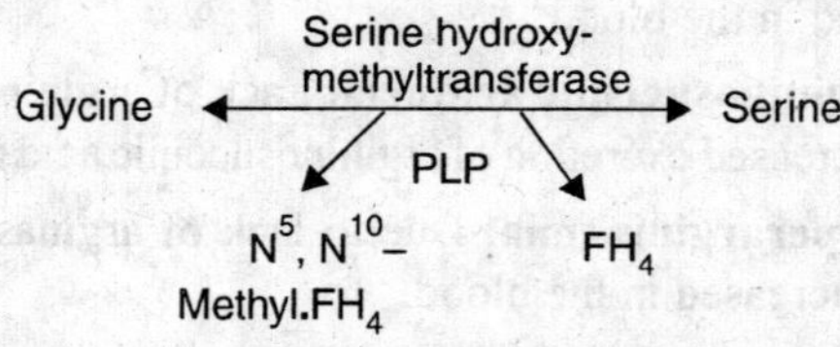

2. Glycine also contributes to C_4, C_5 and N_7 of the purine ring.

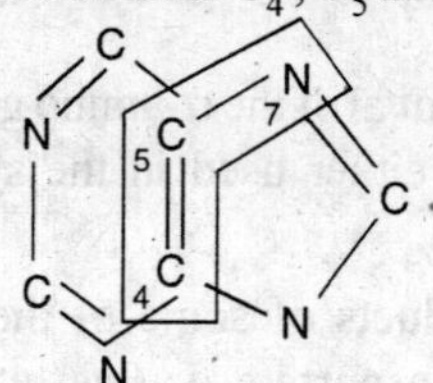

3. It is also a precursor of heme.

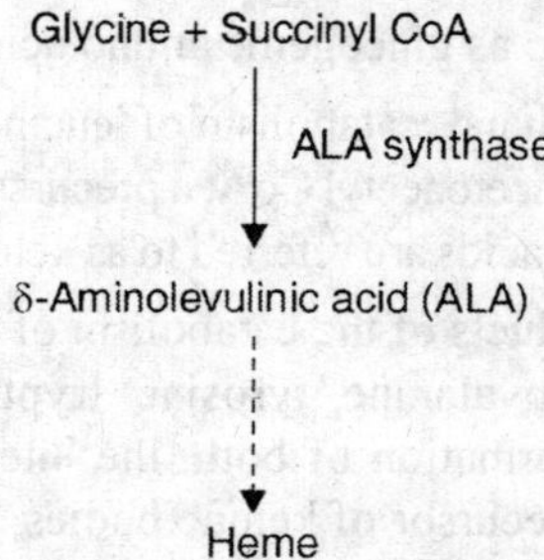

4. Glycine is also used in the biosynthesis of creatine, which is stored as creatinephosphate in the muscle.

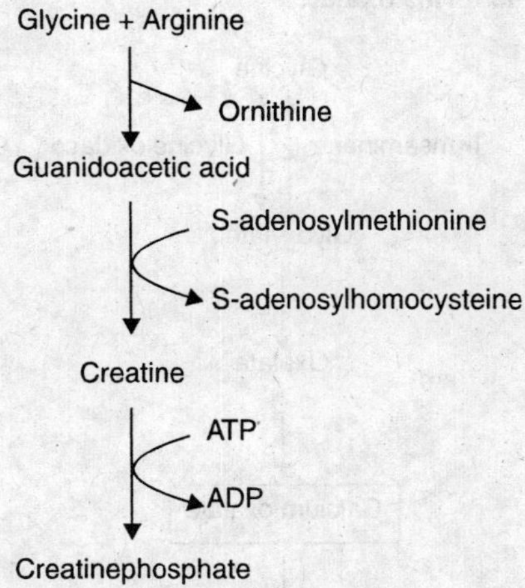

5. Glycine is also used for the synthesis of glutathione.

Glycine + cysteine + glutamic acid

Glutathione (GSH)

6. Glycine is also used in the synthesis of glycholic acid (a bile acid)

Glycine + cholic acid

↓

Glycholic acid

7. Glycine is also used in the detoxification of xenobiotic, *e.g.* benzoic acid

Glycine + Benzoic acid ⟶ Hippuric acid

Q. What is primary hyperoxaluria ?

Ans. Primary hyperoxaluria is an inborn error of glycine metabolism. It is due to the lack of glycine transaminase. Glycine, in turn, results in the formation of oxalate that gets precipitated as calcium oxalate, resulting in the formation of renal stones.

Normally, glycine is oxidised, by glycine oxidase, to

glyoxylate which can again get transaminated by glycine transaminase and forms glycine. When this transaminase is absent glycine forms oxalate.

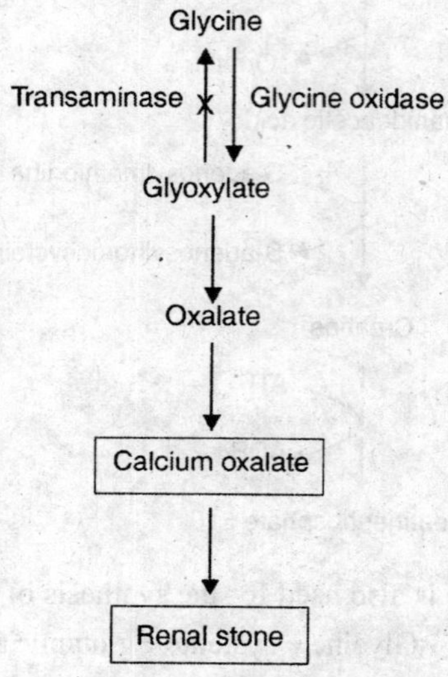

Q. Describe various products which can be synthesized in the body from glutamate ?

Ans. Glutamate is used in the biosynthesis of several compounds in the body –

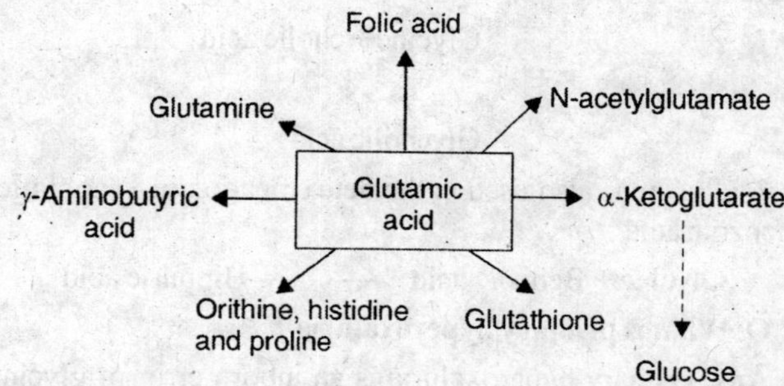

1. Glutamate is converted to glutamine by glutamine synthetase. In the brain, this reaction is important for the removal of ammonia.

Glutamate
NH_3 ↘ | Glutamine synthetase
↓
Glutamine

2. Glutamate is also a precurssor of glutathione

Glutamate + Cysteine + Glycine
↓
Glutathione

3. Glutamate is also convertable to other nonessential aminoacids, such as histidine, proline and ornithine.

4. By transamination, glutamate is converted to α-ketoglutarate which is an intermediate of the Krebs cycle. Hence, glutamate is a glucogenic amino acid.

5. By glutamate decarboxylase, glutamate is also converted to an inhibitory neurotransmitter, *i.e.* γ-aminobutyric acid (GABA).

Glutamate
| Glutamate decarboxylase
CO_2 ↙ |
↓
GABA
(γ-Aminobutyric acid)

6. Glutamate molecules are also incorporated in the biosynthesis of pteroylglutamate (folic acid), by bacteria in the intestinal flora.

7. Glutamate is also used in the synthesis of N-acetylglutamate which is used as a coenzyme by carbamoylphosphate synthetase, in the urea cycle.

Q. What are various metabolic fates of aspartate ?

Ans. Aspartate is a precursor of various compounds –

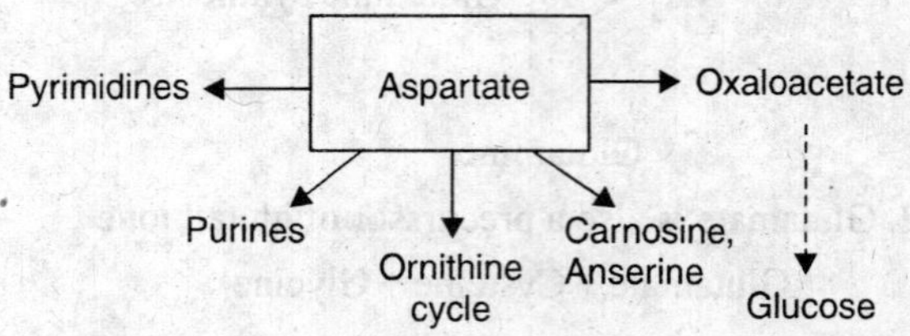

1. Aspartate is incorporated in the biosynthesis of pyrimidines.

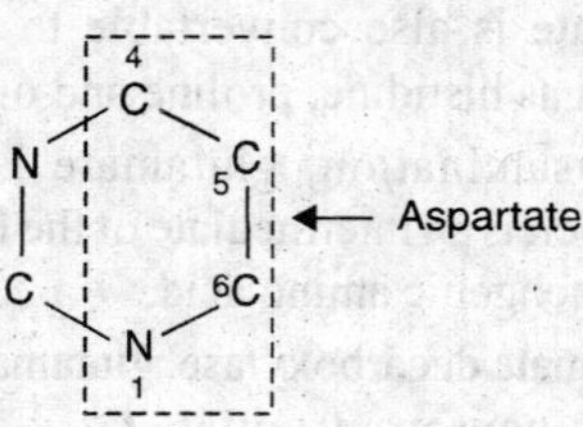

2. It also contributes in the biosynthesis of a purine ring as N – 1.

3. Aspartate also contributes its amino group in the synthesis of urea.

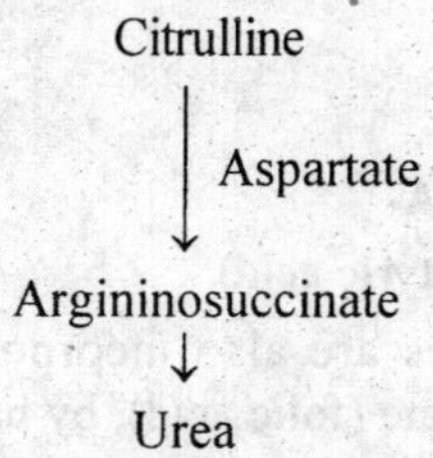

4. By transamination, aspartate is converted to oxaloacetate, an intermediate of the Krebs cycle. Hence, aspartate is a glucogenic amino acid.

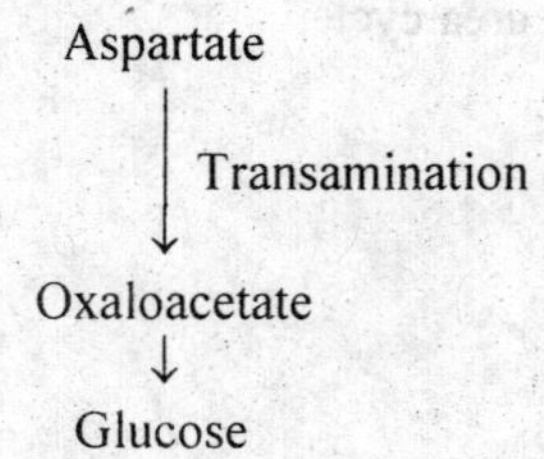

Q. How tyrosine is formed in the body ?

Ans. Tyrosine is formed from phenylalanine. This reaction is catalyzed by phenylalanine hydroxylase.

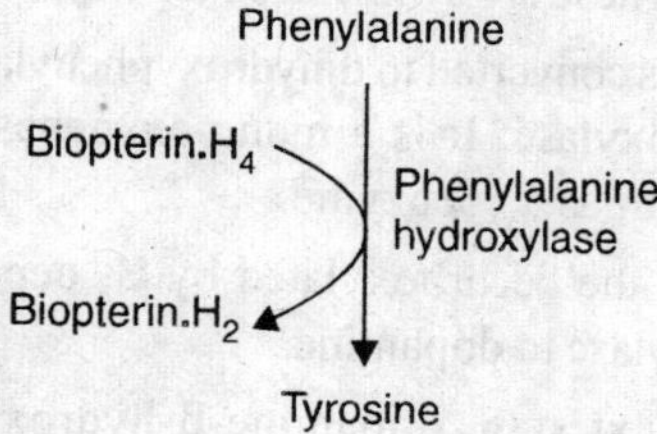

Q. What are the metabolic fates of phenylalanine/tyrosine.

Ans. Phenylalanine is a precursor of tyrosine, thereafter several compounds can be synthesized in the body from tyrosine–

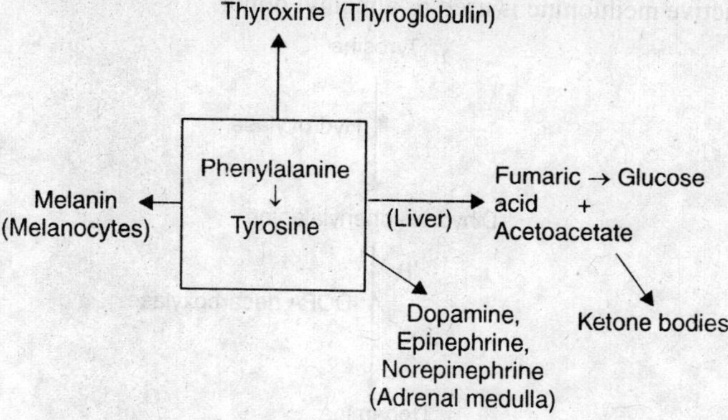

1. Tyrosine is degraded in the liver to fumarate (a precursor of glucose) and acetoacetate (a precursor of ketone bodies). Hence, phenylalanine and tyrosine are both glucogenic as well as ketogenic in nature.

2. In the adrenal medulla, these amino acids are the precursor of catecholamines, *i.e.* dopamine, norepinephrine and epinephrine.

3. In the melanocytes, these amino acids are utilized for the synthesis of melanin, a brown-black pigment.

4. In the thyroid gland, tyrosine residues (in the protein thyroglobulin) are iodinated and used for the synthesis of thyroid hormones.

Q. Describe the process of biosynthesis of norepinephrine and epinephrine.

Ans. Norepinephrine and epinephrine are hormones of the adrenal medulla. These are synthesized from tyrosine as follows :

1. Tyrosine is converted to dihydroxyphenylalanine (DOPA) by tyrosine hydroxylase. It is a mono-oxygenase and requires tetrahydrobiopterin as a coenzyme.

2. DOPA is the decarboxylated by B_6 dependent enzyme DOPA-decarboxylase to dopamine.

3. In the next step, dopamine-β-hydroxylase converts dopamine to norepinephrine. This is a copper containing enzyme and requires vitamin C.

4. Norepinephrine is methylated by phenylethanolamine-N-methyltransferase (PNMT) to epinephrine. In this reactiion, active methionine is used as a methyl donor.

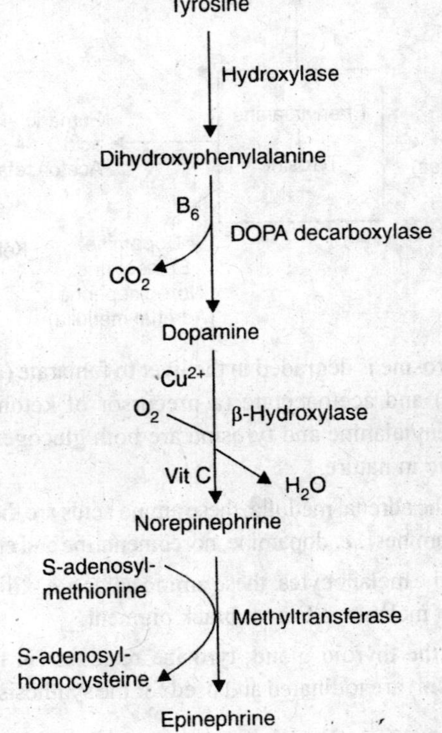

Q. Describe the process of synthesis of melanin from tyrosine.

Ans. Melanin is synthesised from tyrosine in the melanocytes, as follows :

1. Tyrosine is converted to DOPA by Cu^{2+} containing enzyme tyrosinase. Subsequently, DOPA is converted to DOPA quinone.

Melanin is a brown-black pigment present in the skin, hair and retina.

2. By nonenzymatic polymerization, DOPA-quinone is converted to melanin.

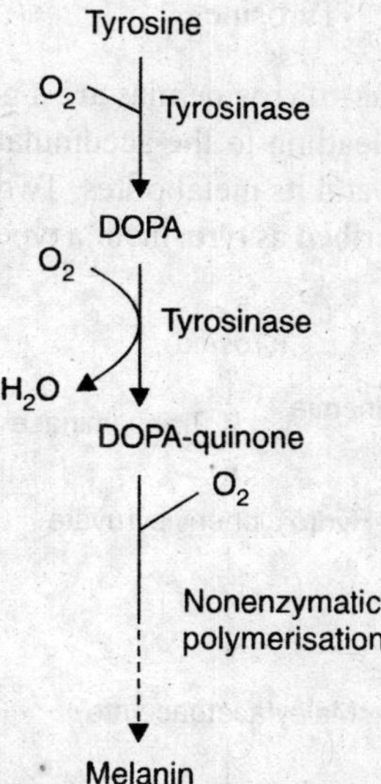

Q. Describe various inborn errors of metabolism of phenylalanine.

Ans. Several defects in phenylalanine metabolism have been shown to exist –

1. **Phenylketonuria**– Phenylketonuria is an inborn error of phenylalanine metabolism due to the deficiency of phenylalanine hydroxylase. It is an autosomal recessive disorder. Since phenylalanine can not be converted to tyrosine, it gets accumulated. Increased level of phenylalanine is metabolized by the alternative pathway and is converted to phenylpyruvate,

phenyllactate and phenylacetate. Due to increased urinary excretion of these phenylketoacids, this condition is referred to as phenylketonuria. These ketoacids may also result in mental retardation and other neurological symptoms.

Diet low in phenylalanine should be given and tyrosine becomes an essential amino acids for such individuals.

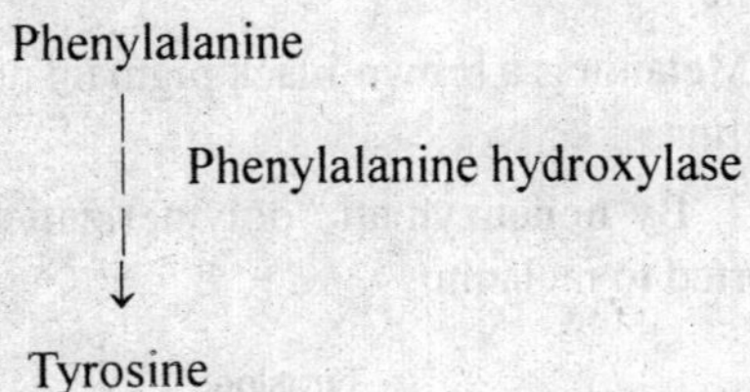

2. **Tyrosinemias**– Tyrosinemias are a group of autosomal recessive disorders, leading to the accumulation and increased excretion of tyrosine and its metabolites. Two common types of tyrosinemias are described as tyrosinemia type-I and tyrosinemia type-II.

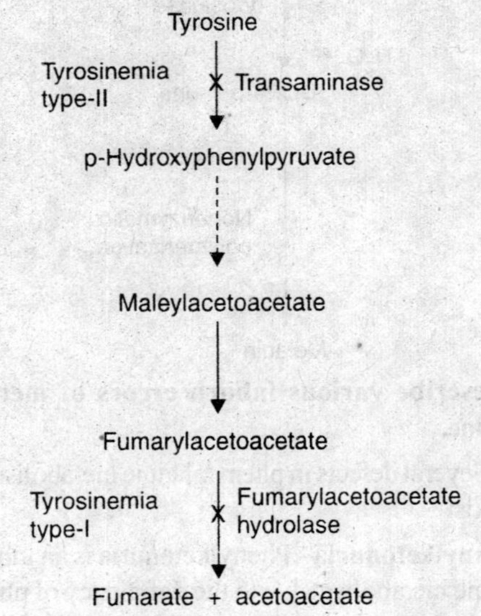

Tyrosinemia type-I is due to the deficiency of fumarylaceto-acetate hydrolase resulting in liver failure and renal tubular dysfunction. It is thus also referred to as hepatorenal tyrosinemia.

Tyrosinemia type-II is due to the deficiency of tyrosine transaminase. Tyrosine gets accumulated and catabolized by the alternative pathway. Accumulation of tyrosine and its metabolites in different tissues, in turn, result in lesions in the eyes and the skin, and cause mental retardation. It is thus also referred to as oculocutaneous tyrosinemia.

3. **Alkaptonuria**– Alkaptonuria is due to lack of homogentisic acid oxidase. Homogentisic acid so formed is not further oxidized but slowly, it gets polymerized to form a dark-black coloured compound, referred to as alkapton. Urine of such individuals turns black, due to auto-oxidation of homogentisic acid. These black colored pigments may also get deposited in the bone, connective tissue and other organs. The condition is referred to as **ochronosis.**

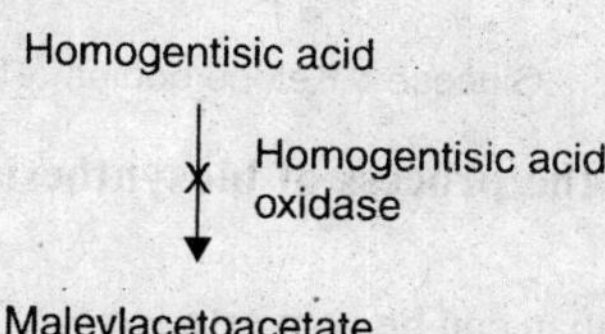

4. Albinism is due to the deficiency of the enzyme tyrosinase. As a result of it melanin is not synthesized, resulting in the formation of white skin and iris. The affected individuals (albinos) are more sensitive to sunlight and prone to burns, skin cancer and photophobia.

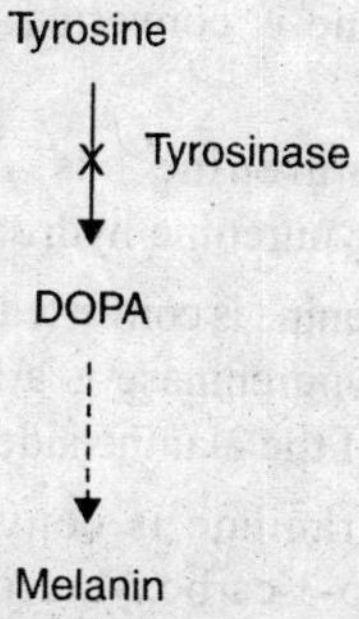

Q. What is Parkinson's disease ?

Ans. Parkinson's disease is associated with a defect in the production of dopamine which is synthesized from phenylalanine/ tyrosine. Lack of its production, in certain parts of the brain results in degenerative condition causing shaking palsy and development of tremors, besides nausea, vomiting, cardiac arrythmias, etc.

Q. What are various metabolic fates of tryptophan ?

Ans. Tryptophan is an essential amino acid. It is both glucogenic as well as ketogenic. It is a precursor of serotonin, NAD and melatonin.

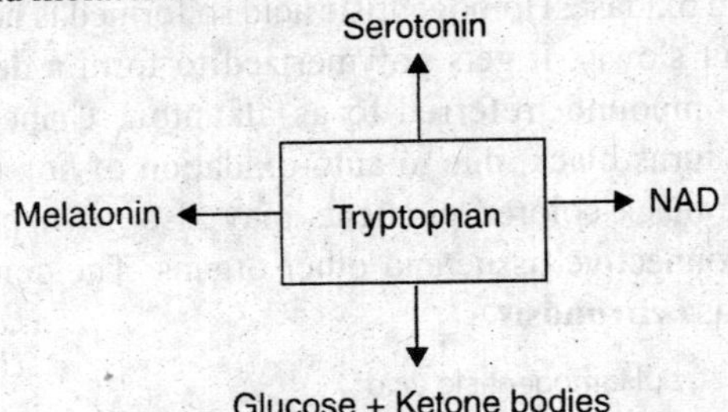

Q. Describe the process of biosynthesis of NAD from tryptophan.

Ans. Tryptophan can be converted to NAD in the liver and some other organs, as follows :

1. Tryptophan is converted to N-formylkynurenine, by the enzyme tryptophan pyrrolase (tryptophan oxygenase).

It is a heme containing enzyme and results in the opening of the indole ring.

2. Formylkynurenine is converted to kynurenine by the enzyme formamidase.

3. N-Formylkynurenine is converted to 3-hydroxykynurenine by kynurenine hydroxylase.

4. 3-Hydroxykynurenine is converted to 3-hydroxyanthranilate by kynureninase. Kynureninase is a B_6 dependent enzyme and results in cleavage of the alanine side chain.

5. 3-Hydroxyanthranilate is converted to an unstable intermediate, *i.e.* 2-amino-3-carboxymuconate-6-semialdehyde. This reaction is catalyzed by 3-hydroxyanthranilate oxidase.

6. Nonenzymatic cyclisation of the intermediate results in the formation of quinolinic acid.

7. Quinolinic acid is converted to nicotinate mononucleotide (NMN) by the enzyme quinolinate phospho-

ribosyltransferase (QPRT). This enzyme transfers a ribonucleotide group from phosphoribosyl pyrophosphate (PRPP).

8. NMN is phosphorylated to NAD, by a kinase.

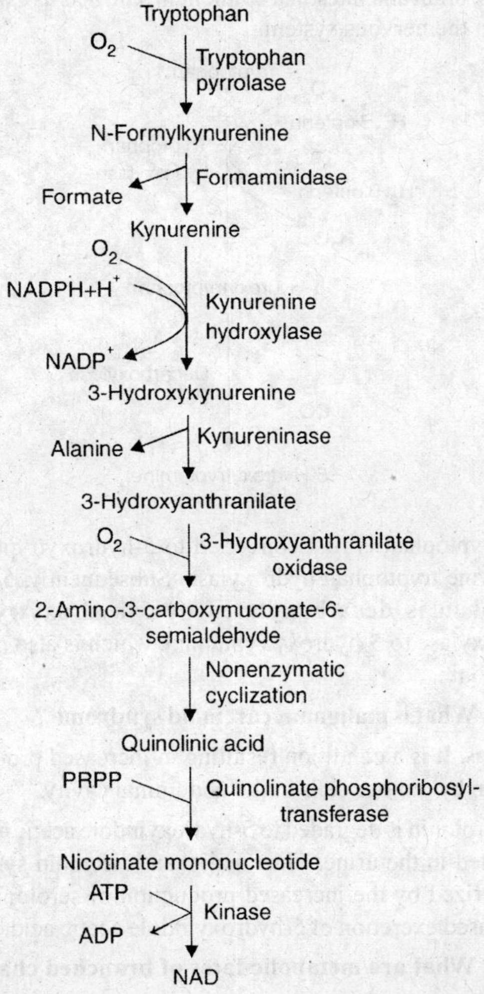

Q. What is tryptophan-niacin conversion ratio ?

Ans. Ratio of the conversion of tryptophan to NAD is referred to as tryptophan-niacin conversion ratio.

Sixty mg of tryptophan can form one mg of niacin, hence tryptophan-niacin conversion ratio is 60 : 1.

Q. How is serotonin synthesized ?

Ans. Serotonin is synthesized from tryptophan in the platelets, brain and intestinal epithelium, and acts as a neurotransmitter in the nervous system.

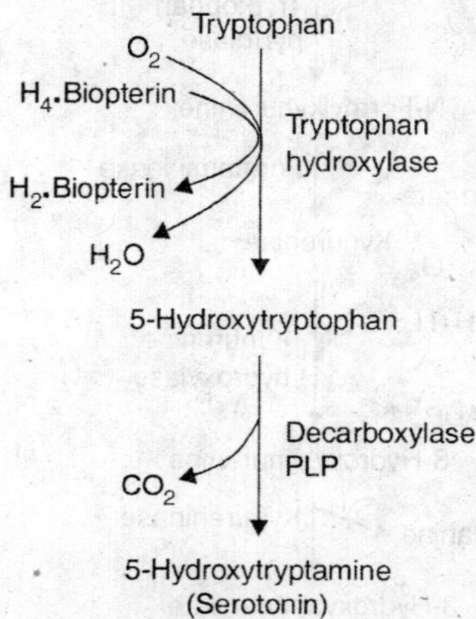

Tryptophan is first converted to 5-hydroxytryptophan by the enzyme tryptophan hydroxylase. Subsequently, 5-hydroxytryptophan is decarboxylated by 5-hydroxytryptophan decarboxylase to 5-hydroxytryptamine which is also referred to as serotonin.

Q. What is malignant carcinoid syndrome ?

Ans. It is a condition resulting in increased production of serotonin by tumour cells in the abdominal cavity.

Serotonin is degraded to 5-hydroxyindole acetic acid which is excreted in the urine. Thus malignant carcinoid syndrome is characterized by the increased production of serotonin as well as increased excretion of 5-hydroxyindole acetic acid.

Q. What are metabolic fates of branched chain amino acids ?

Ans. Valine, leucine and isoleucine are referred to as branched chain amino acids. Out of these valine is purely glucogenic, as the end product of its catabolism is propionyl

CoA. Leucine is purely ketogenic since complete degradation of leucine results in the production of acetyl CoA and acetoacetate where as isoleucine is both glucogenic and ketogenic as the end products of its catabolism are propionyl CoA and acetyl CoA.

The first three reactions in the catabolism of these amino acids are of similar nature and are as follows :

1. Branched chain amino acids undergo transamination by aminotransferases and are converted to branched chain α-ketoacids.

2. The branched chain α-ketoacids are then converted to branched chain acyl CoA thioesters, by the process of oxidative decarboxylation. This reaction is similar to the conversion of pyruvate to acetyl CoA or α-ketoglutarate to succinyl CoA. It is catalyzed by α-ketoacid dehydrogenase complex which requires lipoic acid, TPP, NAD, FAD and CoA.

3. Branched chain acyl CoA thioesters are converted to α, β-unsaturated branched chain acyl CoA thioesters. These thioesters are thereafter oxidized to give different end products.

Valine. The end product of valine oxidation is succinyl CoA which is an intermediate of the Krebs cycle. Hence, valine is a glucogenic amino acid.

Leucine. α, β-Unsaturated branched chain acyl CoA thioester of leucine results in the formation of HMG-CoA which is further catabolized to acetoacetate and acetyl CoA, both of these are ketogenic substrates. Hence, leucine is a purely ketogenic amino acid.

Isoleucine. Degradation of isoleucine results in the formation of propionyl CoA (glucogenic) and acetyl CoA (ketogenic).

Q. What is maple syrup urine disease ?

Ans. It is an inborn error of metabolim, in infants, due to the deficiency of branched chain keto acid dehydrogenase. This, in turn, affects the metabolism of branched chain amino acids, as a result of which the keto acids, of these branched chain amino acids, accumulate. There is increased excretion of these branched chain keto acids in the urine, giving it a typical odour of maple syrup. Hence, this disease is called maple syrup urine disease. It appears in early infancy and is characterized by ketoacidosis, vomiting and convulsions. Reduced intake of branched chain amino acids may improve the condition.

Q. What are various metabolic fates of methionine ?

Ans. Methionine is an essential amino acid. It is glucogenic in nature. It is a precursor of cysteine and is used as a donour of the methyl group, in the form of S-adenosylmethionine (active methionine).

Q. What is active methionine ?

Ans. Active methionine is also referred to as S-adenosylmethionine. It is synthesized from methionine, by the transfer of an adenosyl moiety from ATP, by the enzyme methionine adenosyltransferase.

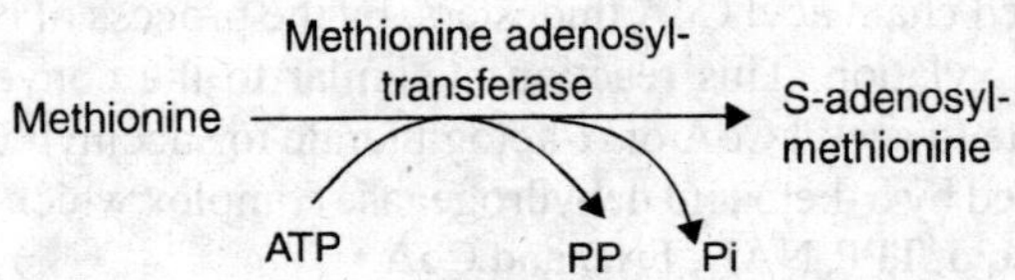

Active methionine acts as a methyl donor and is used for the methylation of various compounds *e.g.* –

1. In the conversion of norepinephrine to epinephrine

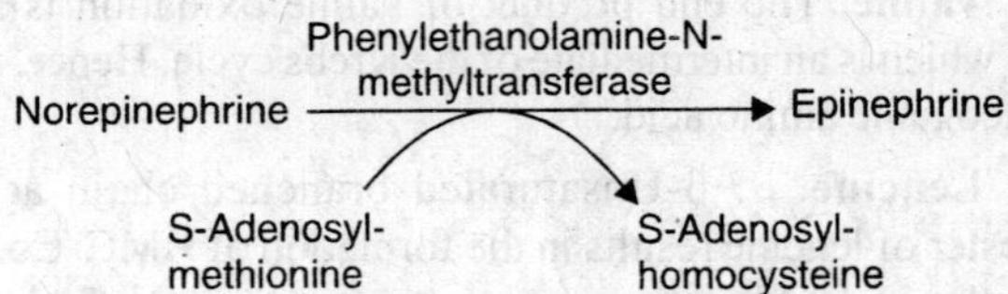

2. Active methione is also used in the biosynthesis of polyamines (spermidine and spermine).

Q. What is hyperhomocysteinemia ?

Ans. Hyperhomocysteinemia is an autosomal recessive disease due to the lack of cystathionine synthase, in the liver. Complete or partial deficiency of the enzyme results in accumulation of homocysteine in the blood (hyperhomocysteinemia) and its increased excretion in the urine (homocysteinuria).

It is characterized by mental retardation, skeletal deformities involving the spine and thorax, seizures and dislocation of the eye lens.

Hyperhomocysteinemia may also result in defective collagen biosynthesis. It may also result in atherogenesis due to the formation of homocysteine thiolactose which, in turn, may affect uptake of LDL by the macrophages, leading to atheromas and calcification of atherosclerotic lesions.

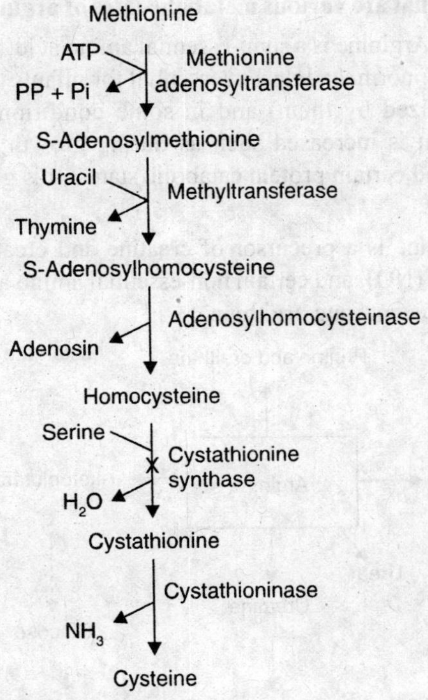

Q. What is cystathioninuria ?

Ans. Cystathioninuria is an inborn error of cysteine metabolism, due to lack of cystathioninase.

This, in turn, results in mental retardation.

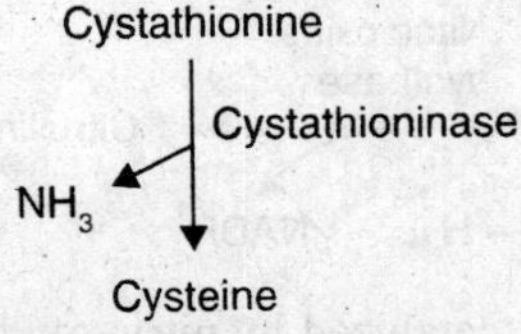

Q. Write note on cystinuria ?

Ans. Cystinuria is a transport disorder due to a defect in the transport of cystine alongwith other basic amino acids such as lysine, arginine and ornithine. It is characterized by amino aciduria, due to increased renal excretion of these amino acids. Due to low solubility of cystine, renal calculi may also be observed.

Q. What are various metabolic fates of arginine ?

Ans. Arginine is a semi-essential amino acid, *i.e.* it is non-essential for normal adults but essential for infants (as it can not be synthesized by them) and in some conditions where its requirement is increased such as during growth, pregnancy, lactation and certain protein catabolic states. It is glycogenic in nature.

Arginine is a precursor of creatine and creatinine, urea, nitric oxide (NO), and certain non-essential amino acids such as proline, glutamate and ornithine.

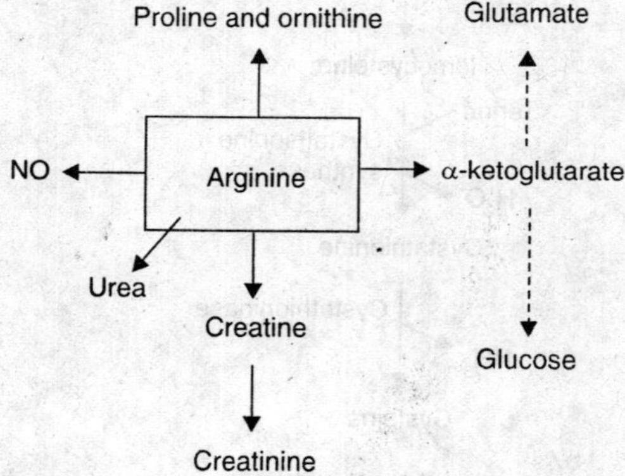

Q. Write a note on nitric oxide.

Ans. Nitric oxide (NO) is a free radical (diatomic gas), synthesized endogenously from the guanidino group of arginine.

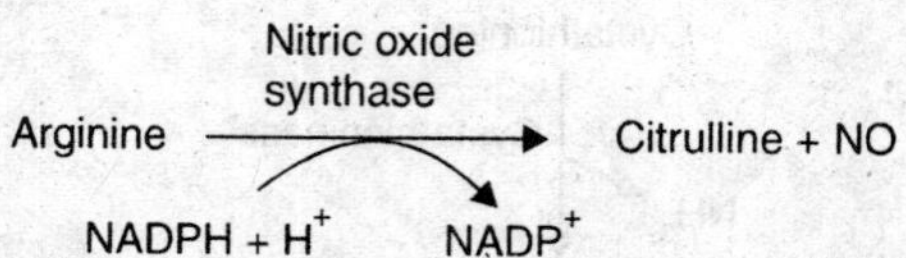

This reaction is catalyzed by nitric oxide synthase. NO is an active agent in the dilatation of the blood vessels.

Q. What are metabolic fates of histidine ?

Ans. 1. Histidine is a semi-essential amino acid. It is a glycogenic amino acid and important in one-carbon metabolism.

2. It is also a precursor of histamine.

Q. What are polyamines ?

Ans. Polyamines are highly cationic molecules that are synthesized before mitosis occurs, *e.g.* putresine, spermidine and spermine. These are essential growth factors required for cell proliferation.

Polyamines are synthesized in the liver from ornithine.

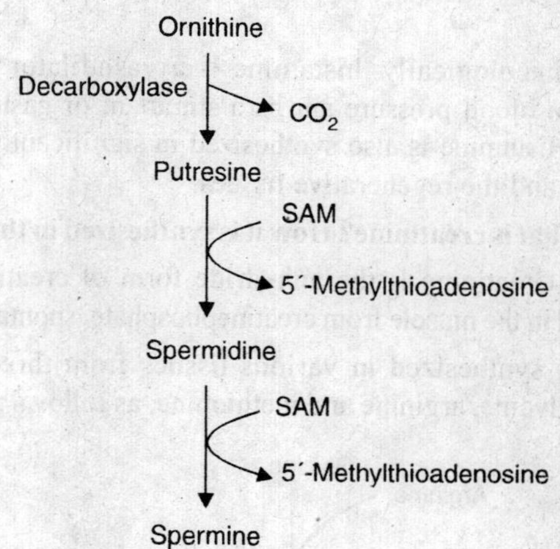

Most of the requirement of polyamines by the body is met by their synthesis, by the microflora of the intestine.

Spermidine is also required in protein synthesis and helps in post-translational modification of the eukaryotic initiation factor-4D (eIF 4-D).

Polyamines also act as sedatives in pharmacological doses.

Q. What is FIGLU ?

Ans. FIGLU (Formiminoglutamate) is formed as an intermediate in histidine degradation. Formimino group from FIGLU is accepted by tetrahydrofolate (FH_4) and glutamate is formed.

In patients with folate deficiency, this reaction is inhibited as formimino group can not be transferred to tetrahydrofolate and FIGLU is excreted in the urine, without being converted to glutamate. Hence, FIGLU excretion test is done to assess folate status of an individual after a test dose of histidine. Increased excretion of FIGLU is diagnostic of folate deficiency.

Q. What is histamine and how it is synthesized ?

Ans. Histamine is synthesized from histidine in a reaction that is catalyzed by histidine decarboxylase. Histamine production is increased in an allergic reaction.

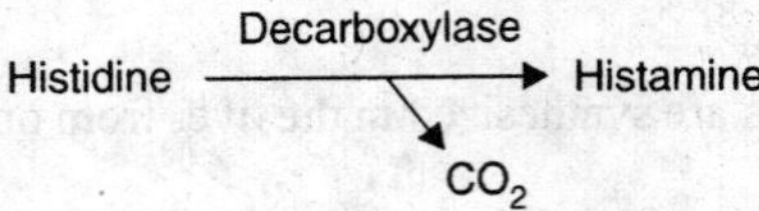

Pharmacologically, histamine is a vasodilator causing reduction in blood pressure and is a stimulant of gastric juice secretion. Histamine is also synthesized in significant amounts in the fetal and the regenerative tissues.

Q. What is creatinine? How it is synthesized in the body ?

Ans. Creatinine is the anhydride form of creatine. It is synthesized in the muscle from creatinephosphate, spontaneously.

Creatine is synthesized in various tissues from three amino acids, *i.e.* glycine, arginine and methionine, as follows :

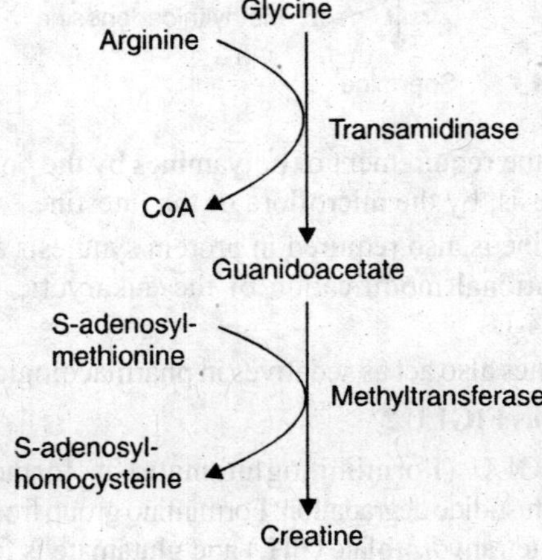

1. In the first step, glycine accepts a guanidino group from arginine and is converted to guanidoacetate. This reaction is catalyzed by arginine-glycine transamidinase.

2. Guanidoacetate accepts a methyl group from S-adenosyl-methionine and thus creatine is formed. This reaction is catalyzed in the liver, by methyltransferase.

3. A high energy phosphate group is transferred from ATP to form creatinephosphate. This reaction is catalyzed by creatine-phosphokinase.

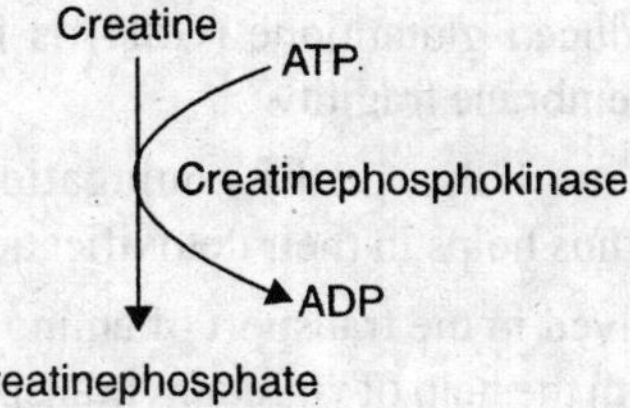

4. Creatine phosphate is the stroage form of energy in cardiac and skeleted muscle.

It is utilized to phosphorylate ADP which is converted to ATP, that is required during muscular contraction.

During phosphorylation of ADP, by creatinephosphate, creatine is released as creatinine. This reaction is referred to as Lohman's reaction. It is a reversible reaction and is catalyzed by creatinephosphokinase (CPK).

Creatinine is a waste product and is excreted in the urine.

Q. What is glutathione? How it is synthesized?

Ans. Glutathione is a tripeptide consisting of glycine, cysteine and glutamate. It is synthesized in various tissues, as follows :

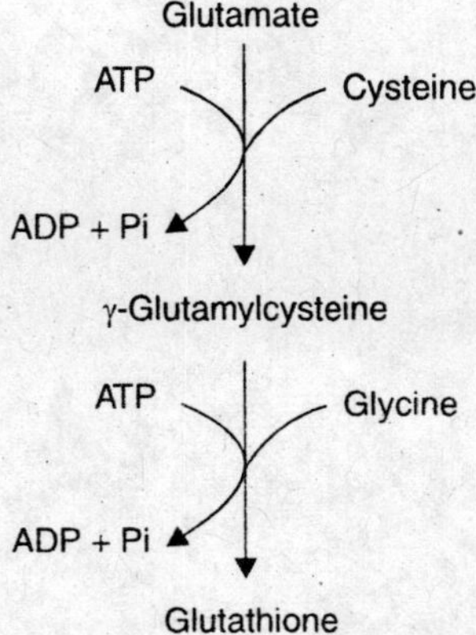

Glutathione (GSH), due to the presence of cysteine, has an

–SH group and thus exists in the oxidized (GSSG) as well as reduced state (GSH).

Q. What is the physiological significance of glutathione ?

Ans. 1. Reduced glutathione (GSH) is important in the maintenance of membrane fragility.

2. Glutathione helps in the conjugation of drugs and xenobiotics, and thus helps in their detoxification.

3. It is involved in the transport of amino acids across the cell membrane with the help of γ-glutamyltranspeptidase (γ-GT).

Integration of Metabolism

Q. Discuss the role of various tissues in fuel metabolism.

Ans. Several metabolic pathways which are concerned with the oxidation of fuel for the production of ATP such as glycolysis, citric acid cycle, fatty acid oxidation, lipogenesis, amino acid oxidation and proteolysis, etc., do not operate in every tissue and at all the times. These various metabolic processes are integrated and have interdependence of several tissues of the body.

Five major organs i.e. liver, adipose tissue, muscle, brain and kidney play a dominant role in fuel metabolism. These tissues contain unique sets of enzymes in such a way that each organ is specialized for the storage, use and generation of specific fuels. These tissues do not function in isolation but are interdependent on each other in which one tissue may provide substrates to another or process the compounds produced by other organs. Metabolites flow between these organs in the well-defined pathways, in which flux also varies with the nutritional state of the animal. Communication between the tissues is mediated by the nervous system, by the availability of circulating substrate and by variations in the levels of plasma hormones. For example, immediately following meal, glucose, amino acids and fatty acids are directly available from the intestine. Glucose passes from the intestinal epithelial cells via the portal vein, to the liver. Amino acids are partially metabolized in the gut before being released into the portal blood. Chylomicrons, containing triacylglycerol, are secreted by the intestinal epithelial cells into the lymphatic and via thoracic duct are delivered to the rest of the body. This integration of energy metabolism is controlled, primarily by insulin and the opposing actions of glucagon and epinephrine.

ROLE OF LIVER

Liver is referred to as central metabolic clearing house of the body as it maintains proper levels of the circulating fuels, for use by the brain, muscle and other tissues. Moreover, all the nutrients absorbed by the intestine, except fatty acids, directly reach liver via portal vein.

Glucose : One of the major functions of the liver is to act as a buffer for blood glucose. It takes up and releases glucose in response to hormones and to the circulation of blood glucose. When the blood glucose concentration is high, such as after a carbohydrate containing meal, liver takes up glucose and converts it to glucose-6-phosphate, by the enzyme glucokinase. Glucose-6-phosphate thereafter has several fates in the liver:

- It can be converted to glucose by glucose-6-phosphatase for transport to the peripheral organs when blood glucose concentration is reduced.
- It can be converted to glycogen, when the demand of the body for glucose is low.
- Glucose-6-phosphate can be converted to acetyl CoA, via glycolysis, which (if not oxidized via citric acid cycle for the generation of ATP) can be used to synthesize fatty acids, ketone bodies and cholesterol.
- It can also be degraded via the pentose phosphate pathway to generate NADPH, which is required for the biosynthesis of fatty acids, cholesterol and other compounds.

Triacylglycerol : Liver can synthesize or degrade triacylglycerol. Under conditions of high metabolic demand, fatty acids are degraded to acetyl CoA and then to ketone bodies. Liver also generates ATP from acetyl CoA and exports ketone bodies to the peripheral tissues, to be used there as fuel.

When the demand for metabolic fuel is low, fatty acids are incorporated into triacylglycerol that are secreted into the blood stream as VLDL, for uptake by adipose tissue.

Amino Acids : The liver degrades amino acids to a variety of metabolic intermediates that can be completely oxidized to CO_2 and H_2O, or converted to glucose or ketone bodies.

Immediately after food, when amino acids are present in high concentration in the blood, their oxidation provides a significant fraction of metabolic energy.

During a fast, glucose is produced from amino acids, which

mostly arise from degradation of muscle protein.

ROLE OF BRAIN

Brain has a very high respiration rate and requires a steady supply of glucose from blood, which is its primary fuel. During prolonged fast, brain gradually switches to ketone bodies.

ROLE OF MUSCLE

Major fuels for muscle are glucose, fatty acids and ketone bodies. Glycogen is converted to glucose-6-phosphate for its entry into glycolysis, for use within the muscle only. Muscle contraction is driven by ATP hydrolysis.

Skeletal muscle : About 30% of the O_2 consumed by the human body is used by skeletal muscle at rest, which is increased by several folds, in response to a heavy work load. In the resting muscle, ATP is used to phosphorylate creatine and is stored as phosphocreatine. During the initial stage of exercise, ATP is regenerated by the reaction of phosphocreatine with ADP. With the increasing intensity of exercise, muscle shift to ATP production via glycolysis and most of the glucose-6-phosphate is then degraded anaerobically.

During exercise, in humans, the supply of endogenous ATP is made available for a few seconds by phosphocreatine after which anaerobic glycolysis generates ATP. Subsequently, after about 90 seconds, the shift from anaerobic to aerobic metabolism occurs.

Cardiac muscle : Heart muscle relies entirely up on aerobic metabolism. Heart can metabolize fatty acids, ketone bodies, glucose, pyruvate and lactate. Fatty acids are used by the resting heart while during heavy work; the heart greatly increases its consumption of glucose.

ROLE OF ADIPOSE TISSUE

Adipose tissue is widely distributed throughout the body but occurs most prominently under skin, in the abdominal cavity and in skeletal muscle. The function of adipose tissue is to store and release fatty acids as needed for fuel. The adipose tissue of a normal 70 Kg man contains about 15 Kg of fat, which is sufficient to meet the demand of energy for nearly 3 months.

For storage, adipose tissue obtains most of its fatty acids

from circulating lipoproteins. They are activated and stored as triacylglycerol. At the time of metabolic need, adipocytes hydrolyze triacylglycerol to fatty acids and glycerol, by the action of the hormone-sensitive lipase. Fatty acids may be either reutilized for the synthesis of triacylglycerol or if glycerol-3-phosphate is in short supply, they are released into the blood stream.

ROLE OF KIDNEY

Kidney filters urea and other waste products from the blood. It also maintains pH of the blood. It also retains important metabolites, such as glucose. Kidney also generates glucose by gluconeogenesis, particularly during starvation.

The complexity of the mechanisms that regulates mammalian fuel metabolism permits the body to respond efficiently to the changing energy needs, in response to the changes in the availability of various fuels, maintaining reasonable balance between energy inflow and output, such as during starvation or disorders of fuel metabolism, like diabetes and obesity.

Q. Briefly describe the process of fuel metabolism during starvation.

Ans. After a meal is digested, nutrients are catabolized to small units, for absorption by the intestinal mucosa, from where they pass through the circulation, to the rest of the body. Glucose uptake and glycogen and fatty acid biosynthesis are stimulated by insulin whose concentration in the blood increases in response to high blood glucose concentration.

As tissues take up and metabolize glucose, blood glucose concentration drops, thereby causing the release of glucagon by the α cells of the pancreas. This in turn stimulates glycogen break down and release of glucose from the liver. It also promotes gluconeogenesis from amino acids and lactate. After an overnight fast, combination of increased glucagon and decreased insulin secretion promotes mobilization of fatty acids from adipose tissue. The diminished insulin concentration also inhibits glucose uptake by the muscle. Muscle therefore; switch from glucose to fatty acids metabolism for energy production.

When the fast is prolonged, liver glycogen store becomes depleted and the rate of gluconeogenesis increases. After several days of starvation, liver directs acetyl CoA, which is derived from fatty acids, to the synthesis of ketone bodies that are released into blood, and brain gradually adapts to the uses of ketone bodies as fuel.

Q. Discuss the role of liver in the integration of metabolism.

Ans. Liver is referred to as central metabolic clearing house of the body as it maintains proper levels of the circulating fuels, for use by the brain, muscle and other tissues. Moreover, all the nutrients absorbed by the intestine, except fatty acids, directly reach liver via portal vein.

Glucose : One of the major functions of the liver is to act as a buffer for blood glucose. It takes up and releases glucose in response to hormones and to the circulation of blood glucose. When the blood glucose concentration is high, such as after a carbohydrate containing meal, liver takes up glucose and converts it to glucose-6-phosphate, by the enzyme glucokinase. Glucose-6-phosphate thereafter has several fates in the liver:

- It can be converted to glucose by glucose-6-phosphatase for transport to the peripheral organs when blood glucose concentration is reduced.
- It can be converted to glycogen, when the demand of the body for glucose is low.
- Glucose-6-phosphate can be converted to acetyl CoA, via glycolysis, which (if not oxidized via citric acid cycle for the generation of ATP) can be used to synthesize fatty acids, ketone bodies and cholesterol.
- It can also be degraded via the pentose phosphate pathway to generate NADPH, which is required for the biosynthesis of fatty acids, cholesterol and other compounds.

Triacylglycerol : Liver can synthesize or degrade triacylglycerol. Under conditions of high metabolic demand, fatty acids are degraded to acetyl CoA and then to ketone bodies. Liver also generates ATP from acetyl CoA and exports ketone bodies to the peripheral tissues, to be used there as fuel.

When the demand for metabolic fuel is low, fatty acids are

incorporated into triacylglycerol that are secreted into the blood stream as VLDL, for uptake by adipose tissue.

Amino Acids : The liver degrades amino acids to a variety of metabolic intermediates that can be completely oxidized to CO_2 and H_2O, or converted to glucose or ketone bodies.

Immediately after food, when amino acids are present in high concentration in the blood, their oxidation provides a significant fraction of metabolic energy.

During a fast, glucose is produced from amino acids, which mostly arise from degradation of muscle protein.

8

Vitamins

Q. What are vitamins ?

Ans. Vitamins are organic constituents of food which are essential for life and well-being. Normally, they are not synthesized in the body and have to be supplied in the diet, hence vitamins are also called as "accessory food factors" .

These were discovered as vital substances for the body, most of which were amines and considered to be dietary essential. Now-a-days, it is known that some of the vitamins can also be synthesized in the body though not in required concentration, *e.g.* vitamin D, niacin, etc.

Q. Name fat soluble vitamins.

Ans. Fat soluble vitamins are –

Vitamins A, D, E and K.

Q. Describe the structure of vitamin A.

Ans. Vitamin A has a six-membered ring known as β– ionine ring. This ring is attached with a long hydrocarbon side chain, normally with a primary alcoholic group. This is called vitamin A alcohol. It is also referred to as retinol.

The alcohol (retinol) is interconvertible to aldehyde (retinal, also called as retinene) or acid (retinoic acid).

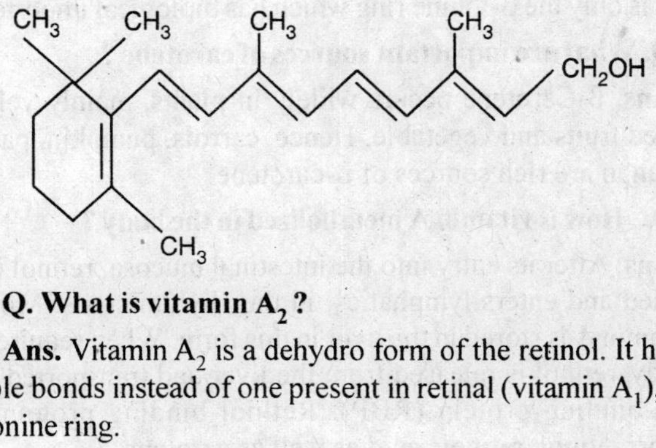

CH_3 CH_3 CH_3 CH_3

CH_2OH

CH_3

Q. What is vitamin A_2 ?

Ans. Vitamin A_2 is a dehydro form of the retinol. It has two double bonds instead of one present in retinal (vitamin A_1), in the β– ionine ring.

Q. How is vitamin A absorbed ?

Ans. Vitamin A is absorbed in the intestine to the extent of about 50-60% and is released into the lacteals. It is absorbed in its esterified form (as retinyl ester) with bile and is hydrolyzed to free retinal in the intestinal lumen.

Q. What is provitamin A ?

Ans. Provitamin A is the precursor of vitamin A. Such compounds are referred to as carotenoids. These are found in plants and have two ionine rings.

Depending upon the presence of the second ionine ring, these are designated as α, β or γ carotenes.

β-carotene has highest vitamin A activity amongst all the carotenoids.

Q. How carotenoids are converted to vitamin A ?

Ans. β-Carotene is splitted at the centre of the hydrocarbon chain within the intestinal wall, by the enzyme retinene reductase, in the presence of molecular oxygen and bile salts.

$$\beta\text{-Carotene} \xrightarrow[\text{Retinene reductase}]{O_2,\ \text{Bile salts}} 2\ \text{Vitamin A}$$

A molecule of β-carotene gives rise to two molecules of vitamin A since it has two β-ionine rings whereas α and γ carotenes, give rise to one molecule of vitamin A because they contain only one β-ionine ring.

It is only the β-ionine ring which has biological importance.

Q. What are important sources of carotene ?

Ans. β-Carotene occurs widely in plants, mainly yellow-coloured fruits and vegetable. Hence, carrots, pumpkin, papaya and mango are rich sources of β-carotene.

Q. How is vitamin A metabolized in the body ?

Ans. After its entry into the intestinal mucosa, retinol is re-esterified and enters lymphatics, in the chylomicrons. Most of the vitamin A is stored in the liver in this form. When required by the body, retinol is released from the liver and transported with retinol-binding protein (RBP). Retinol binding protein can transport retinol, retinoic acid as well as carotene.

Small amount of the retinoic acid is also transported in the blood with albumin. When retinal is required by the target cells, it is released from the protein. Retinol is converted to retinal in the retina for its role in the vision cycle or to retinoic acid in the epithelial tissues for growth and epithelial functions.

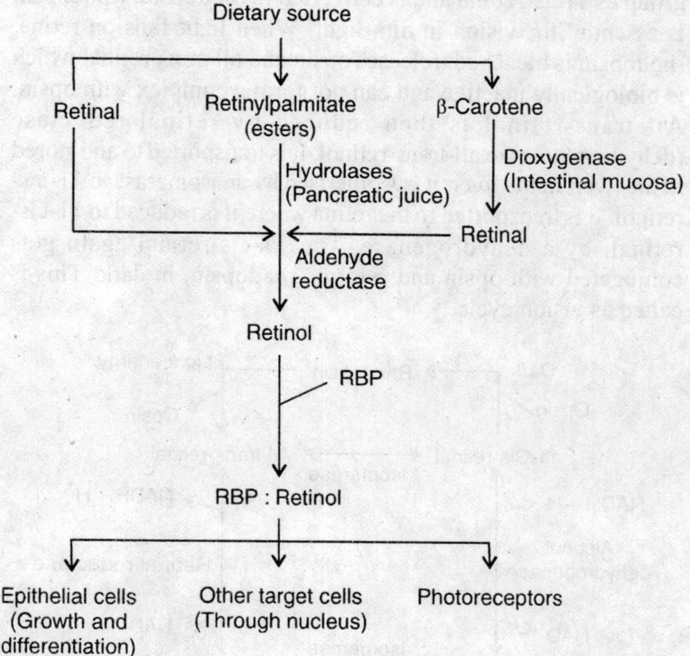

Q. Do all three forms of vitamin A, *i.e.* retinal, retinol and retinoic acid exhibit vitamin A activity ?

Ans. Yes, all the three forms, *i.e.*, retinal, retinol and retinoic acid exhibit vitamin A activity.

Q. What are the biological functions of vitamin A ?

Ans. Important biological functions of vitamin A are :

(*i*) It is essential in vision cycle.

(*ii*) Vitamin A deficiency impairs growth and development.

(*iii*) It is required for the synthesis of glycosaminoglycans (mucopolysaccharides) and proteoglycans (glycoproteins).

(*iv*) Vitamin A deficiency leads to a reduction in gluconeogenesis and cholesterol synthesis.

(*v*) It is also essential for immune functions.

Q. Describe the role of vitamin A in vision cycle.

Ans. Vitamin A (retinal) is essential for vision, both for the rods as well as cones, in the retina.

Rod cells contain a protein called opsin. It has a prosthetic group as 11cis-retinal and is converted to rhodopsin. Rhodopsin is essential for vision in dim-light. When light falls on retina, rhodopsin is bleached, releases opsin and all trans-retinal which is biologically inactive and can not form a complex with opsin. All trans-retinal is then reduced by retinal reductase (dehydrogenase) to all-trans-retinol. It is transported to and stored in the liver. In the liver, it is isomerised by an isomerase to 11-cis-retinol. It is transported to the retina where it is reduced to 11-cis-retinal, by a dehydrogenase. The 11-cis-retinal again gets conjugated with opsin and reforms rhodopsin, in dark. This is called as vision cycle.

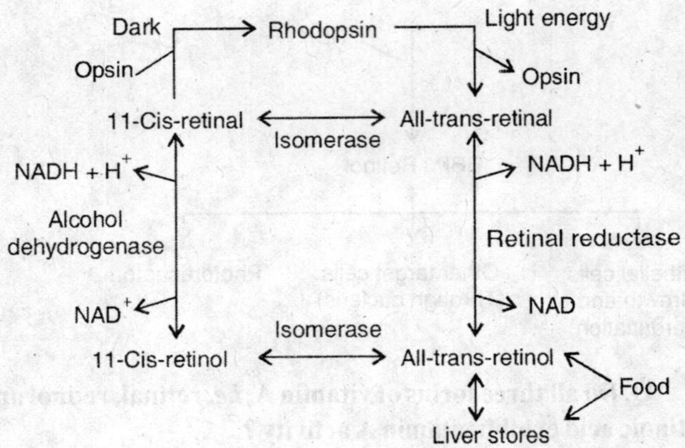

Q. What is the role of vitamin A in cone cells ?

Ans. Cone cells contain three different pigments which are sensitive to primary colours, *i.e.* porphyropsin (for red), iodopsin (for green) and cyanopsin (for blue). The prosthetic group in these three chromopigments (or chromopsins) is 11-cis-retinal.

Q. Name important deficiency diseases due to vitamin A.

Ans. (*i*) Vitamin A deficiency results in xerophthalmia, *i.e.* drying of conjunctiva and cornea with its various features that include –

– Night blindness

– Conjunctival xerosis

– Bitot's spot

– Corneal xerosis/corneal ulceration

– Corneal scar

(*ii*) It also affects osteoblastic activity and causes defective resorption of bone. Vitamin A thus affects bone growth.

(*iii*) Vitamin A deficiency also results in dry and rough skin, due to xerosis and keratinization of the mucous membrane.

(*iv*) It also affects reproductive functions such as degeneration of testes, abortions, malformation of foetus, etc.

Q. What is xerophthalmia ?

Ans. Gradual deficiency of vitamin A results in a condition called xerophthalmia. The disease slowly manifests as night blindness, conjunctival xerosis, Bitot's spot, corneal xerosis and corneal ulceration/keratomalacia.

Q. What are Bitot's spots ?

Ans. Bitot's spot is the formation of a spot in the conjunctiva, due to the accumulation of foamy-cheese like material, as a result of vitamin A deficiency. Though these spots differ in size, shape and location, they have similar appearance.

Q. What is keratomalacia ?

Ans. Prolonged deficiency of vitamin A may lead to corneal ulceration and wasting of the cornea. This is called keratomalacia. Keratomalacia thus indicates destruction of a part or complete corneal stroma, causing a permanent structural alterations.

Q. What are different sources of vitamin A ?

Ans. Vitamin A is obtained from animal sources. Liver oil, egg, milk and milk products are rich sources of vitamin A.

On the other hand, plant sources such as papaya, carrots, mango and green-leafy vegetables are good sources of β–carotene.

Q. What is the daily requirement of vitamin A ?

Ans. Average daily requirement of vitamin A is 2000 to 3000 I.U. (750 to 1000 μg).

Daily requirement of β-carotene is about 4000 to 6000 I.U. (2000 μg).

One I.U. of vitamin A is the activity present in 0.3 μg of retinol or 0.6 μg of β-carotene.

Q. Is vitamin A toxic ?

Ans. Hypervitaminosis A may result in several toxic symptoms since vitamin A is a fat soluble vitamin and is stored in the liver. Important toxicity symptoms include :

1. Dry and rough skin
2. Scaly dermatitis
3. Hepatomegaly
4. Headache, nausea, vomiting and loss of appetite.

Q. Describe chemical nature of vitamin D.

Ans. Vitamin D is a derivative of steroid nucleus and is called sterol, due to the presence of an –OH group. It is also called as anti-rickets factor, sunshine vitamin or calciferol.

It exists in nature in several forms, two of which are prominent. These are referred to as vitamins D_2 and D_3.

Vitamin D_2 is called ergocalciferol. It is of plant origin and is obtained from ergot plant.

Vitamin D_3 is called cholecalciferol. It is of animal origin and is obtained, mainly, from fish oil. It can also be synthesized in the body from its precursor, *i.e.* 7-dehydrocholesterol, by ultraviolet irradiation in the skin.

Q. How vitamin D is absorbed and metabolized in the body ?

Ans. Dietary vitamin D is absorbed in the jejunum. It binds with vitamin D-binding protein in the plasma and is transported to the liver. Thereafter, it is sequentially converted to its active form, *i.e.* 1, 25-dihydroxycholecalciferol also referred to as calcitriol, in the liver and the kidney.

Q. What is provitamin D ?

Ans. 7-Dehydrocholesterol (a derivative of cholesterol), is a precursor of vitamin D_3 in animals and is called provitamin D. It is converted to vitamin D_3 by the action of UV light (from sunlight), in the skin. This process is also called photobio-genesis (photobiosynthesis) of vitamin D.

Ergosterol is a provitamin D in plants.

Q. What is active vitamin D ?

Ans. 1, 25-Dihydroxycholecalciferol is called active vitamin D. It is also called as calcitriol because it has three –OH groups.

Q. How does activation of vitamin D occurs ?

Ans. Vitamin D, both from the dietary source as well as after its synthesis in the skin, is converted to 25–OH– cholecalciferol in the liver. This reaction is catalyzed by 25-hydroxylase.

25–OH– cholecalciferol is transported to the kidney and is converted to 1, 25-dihydroxycholecalciferol, by 1–α–hydroxylase.

1–α–hydroxylase activity is regulated by the parathyroid hormone.

Q. What are important biological functions of vitamin D ?

Ans. Most important biological function of vitamin D is regulation of serum calcium and phosphorous through its action in the intestine, bone and the kidney :

In the intestine vitamin D stimulates the synthesis of calcium-binding protein which, in turn, increases the intestinal absorption of calcium.

In the bone it stimulates osteoclastic activity and induces the synthesis of alkaline phosphatase resulting in the demineralization of bone

In the kidneys vitamin D increases tubular reabsorption of calcium and excretion of phosphorus.

Hence, an overall effect of active vitamin D is to normalize serum calcium and phosphorus levels.

Q. What are the diseases due to vitamin D deficiency ?

Ans. Deficiency of vitamin D leads to rickets in children and osteomalacia in adults.

Ricket is a result of the deficiency of vitamin D in children. Its main symptoms include enlarged wrist, ankles and bowed-legs. Characteristics features include bowed-legs and knock-knees, hence an infant can be clinically diagnosed while walking. Rickets is commonly observed in children who are not fully exposed to sunlight, and live in cold and dingy areas.

Osteomalacia is due to vitamin D deficiency in adults. It is commonly observed in females, during the reproductive phase of their life, and mainly affects pelvic bone. Pelvic bone shows characteristic deformity due to demineralization which, in turn, complicates parturition. It is also commonly observed in women who are not fully exposed to sunshine and remain in purda (wear Burka), due to their social customs.

Q. How can you confirm diagnosis of rickets/ osteomalacia, biochemically ?

Ans. Patients with rickets or osteomalacia often show decreased serum calcium and/or phosphorus along with increased alkaline phosphatase activity. Serum calcium x phosphorus product is generally below 30 in the affected individuals, compared to 40 or more in normal individuals. These subjects also show increased urinary excretion of phosphorus.

Q. What is vitamin D-resistant ricket ?

Ans. As active vitamin D is synthesized in the body after its hydroxylation in the liver and the kidney, and the final step in its activation is catalyzed by 1–α–hydroxylase in the renal tubules, inherited deficiency of 1–α–hydroxylase or renal tubular damage will not result in the formation of active vitamin D. This, in turn, results in its deficiency which can not be cured by the supplementation of Vitamin D. Such a disorder is called as vitamin D-resistant ricket. These children can only be treated by calcitriol (instead of vitamin D).

Q. What are important sources of vitamin D ?

Ans. Exposure to sunlight is the best source of vitamin D, as it can be synthesized in the body.

Rich sources are milk and milk products, fish oil, liver oil and egg yolk.

Q. What is daily requirement of vitamin D ?

Ans. Average daily requirement of vitamin D is 200 – 400 I.U. (5–10 μg).

Q. What is hypervitaminosis D ?

Ans. Intake of extremely large doses of vitamin D may lead to several toxic symptoms and is called as hypervitaminosis D. It may result in hypercalcemia, hyperphosphatemia besides anorexia, nausea, vomiting and diarrhea. It may also result in metastatic calcification of soft tissues including kidney, myocardium, pancreas and uterus.

Q. What is the chemical nature of vitamin E ?

Ans. Several organic compounds, grouped as tocopherols (such as α, β, γ, etc.) have vitamin E activity. d-α-Tocopherol is a natural vitamin E while *dl*-α-tocopherol is a synthetic compound and is available commercially.

α-Tocopherol is an alcohol which has a phytol side chain that is attached to a trimethylhydroquinone ring.

Q. How is vitamin E absorbed and metabolized ?

Ans. Vitamin E is readily absorbed from the gastrointestinal tract with dietary fat, and is transported to the liver and peripheral tissues in the lipoproteins. High dose of vitamin C decreases its absorption.

Level of vitamins E, in lipoproteins, is also affected by various pro- and antioxidants, selenium and sulphur containing amino acids.

Q. Describe biological functions of vitamin E ?

Ans. Most important role of vitamin E is an antioxidant as it has the ability to detoxify free radicals.

Q. Describe effects of vitamin E deficiency.

Ans. Vitamin E deficiency may lead to acne, anemia and infections. Its deficiency is commonly seen in patients with abetalipoproteinemia.

Deficiency of vitamin E is also seen in fat malabsorption. Such patients show signs of progressive neuropathy and retinopathy. Several reproductive defects, such as sterility, degeneration of testes and regression in fetal resorption, have also been reported to occurs in various animal species.

Q. What are free radicals ?

Ans. Free radicals are the powerful oxidizing agents which can oxidize various macromolecules such as proteins, lipids and nucleic acids. Some of the free radicals, *e.g.* hydroperoxides are very unstable and rapidly disintegrate into potentially toxic fragments. This in turn may also lead to the deposition and accumulation of lipofuschin (age-pigments) in the neurons and the heart cells.

Q. What are the important sources of vitamin E ?

Ans. Tocopherols are present in green-leafy vegetables, milk and milk products, and egg yolk. Vegetable oils, such as corn oil, cotton-seed oil, are rich sources of vitamin E.

Q. What is the daily requirement of vitamin E ?

Ans. Average daily requirement of vitamin E in an adult man is nearly 15 I.U. It however, varies with the intake of polyunsaturated fatty acids (PUFA), exposure to stress, and polluted environment. Normally, 0.4 mg of vitamin E is required per g of PUFA.

Q. Describe chemical nature of vitamin K and its biological significance.

Ans. Various compounds related to methylnapthoquinone have a role in blood clotting and are described as vitamin K. Commonly found substances include vitamin K_1, K_2 and K_3.

Vitamin K_1 is also called phyloquinone. It is found in plants. It has a phytyl side chain.

Vitamin K_2 is also called menaquinone or farnequinone. It is synthesized by the intestinal bacteria and has a farnesyl side chain.

Both K_1 and K_2 are fat soluble.

Vitamin K_3 or menadione is a synthetic compound, available commercially. It is water soluble and is readily reabsorbed on parenteral administration.

Q. How is vitamin K absorbed ?

Ans. Vitamin K is absorbed with dietary fat, in the presence of bile salts. It is utilized in the liver.

Q. Describe important biological functions of vitamin K.

Ans. Vitamin K is also called as coagulation factor since it is essential for the synthesis and activation of some of the blood clotting factors such as prothrombin, factor VII, IX and X.

These factors are synthesized in the body as their inactive precursors and are rich in glutamic acid which undergoes post-translational modifications. Glutamic acid residues of these blood coagulation factors are carboxylated and converted to γ-carboxy-glutamyl-residues, forming a site for the binding of Ca^{2+}. For the conversion of glutamyl-residues to carboxy-glutamyl- residues, vitamin K is required. During this conversion vitamin K, in turn, is oxidized to vitamin K-epoxide and is regenerated by epoxide reductase, in the liver.

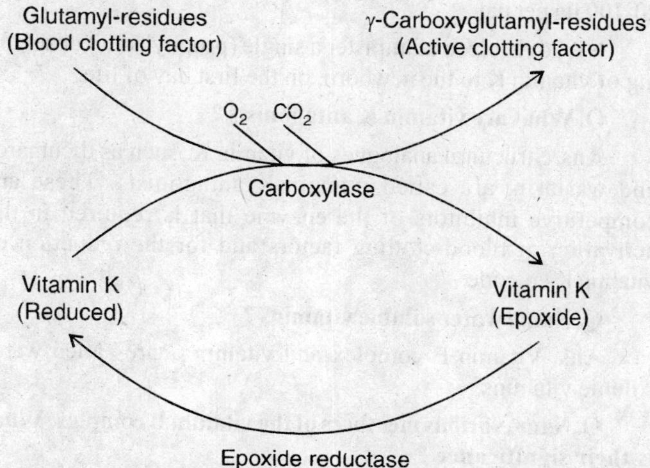

Q. What is the biological importance of γ-carboxylation of glutamate ?

Ans. γ–Carboxylation of the glutamyl-residues of the protein (blood clotting factor) such as prothrombin, factor VII, factor IX and factor X, in turn, helps in activation of the protein and thus in the clotting of blood. It is a post-translational modification of the protein.

Q. Describe deficiency symptoms of vitamin K.

Ans. Vitamin K deficiency generally results in delayed prothrombin time and hemorrhagic tendency. It is commonly seen in patients with fat malabsorption (sprue), obstructive jaundice, advanced liver disease (liver carcinoma, cirrhosis, etc.) or on anticoagulant therapy. Vitamin K deficiency is also seen in a newborn due to sterile intestine, particularly in a premature infant.

Common deficiency symptoms include cutaneous and intramuscular hemorrhages with bluish-red coloration in different parts of the body.

Q. Name some important sources of vitamin K.

Ans. Green-leafy vegetables are rich sources of vitamin K. Besides green-leafy vegetables, fish, liver and muscle meat are also good sources of Vitamin K.

Q. What is daily requirement of vitamin K ?

Ans. Average, recommended dietary allowance (RDA) is 20-100 μg per day.

It is advisible to administer a single (prophylactic) dose of 1 mg of vitamin K to the newborn, on the first day of life.

Q. What are vitamin K antagonists ?

Ans. Structural analogues of vitamin K, such as dicumarol and warfarin, are called vitamin K antagonists. These are competitive inhibitors of the enzyme that is required in the activation of blood-clotting factors and for the reduction of vitamin K epoxide.

Q. Name water soluble vitamins ?

Ans. Vitamin B complex and vitamin C are called water soluble vitamins.

Q. Name various members of the vitamin B complex. What is their significance ?

Ans. Various vitamins of the B complex group are :

Thiamin (vitamin B_1)

Riboflavin (vitamin B_2)

Niacin (vitamin B_3)

Pyridoxine (vitamin B_6)

Pantothenic acid

Biotin

Folic acid

Choline

Lipoic acid, and

Cyanocobalamin (vitamin B_{12})

These vitamins are mainly used as coenzymes and form components of the enzyme-prosthetic group complex.

Q. Describe structural features of thiamin.

Ans. A molecule of thiamin has a pyrimidine nucleus which is linked to a thiozole ring.

Q. What is the significance of thiamin for the body ?

Ans. Thiamin, as thiamine pyrophosphate (TPP), acts as a coenzyme in oxidative decarboxylation reactions, such as –

Pyruvate ⟶ Acetyl CoA

α-Ketoglutarate ⟶ Succinyl CoA

TPP is also used as a coenzyme in transketolase reaction, in the hexosemonophosphate shunt (HMP shunt).

Q. Name deficiency disorders of thiamin.

Ans. Thiamin deficiency leads to beri-beri, hence this vitamin is also called as antiberi-beri vitamin.

Q. What is beri-beri ?

Ans. Beri-beri is a disease due to the deficiency of thiamin (vitamin B_1). It affects several tissues and accordingly, beri-beri can be classified into wet beri-beri and dry beri-beri.

Wet beri-beri : It affects cardiovascular system, resulting in dilated heart and edema.

Dry beri-beri : It mainly affects central nervous system and causes peripheral neuropathy with myelin degeneration leading to foot-drop, wrist-drop and sensory changes.

Due to its role in the nervous system, thiamin is also called aneurine.

Deficiency of thiamin commonly occurs in population consuming polished-rice or raw-fish. Its deficiency is also observed in chronic alcoholics. Deficiency symptoms include nausea, vomiting, muscle weakness and tenderness of the calf-muscles.

Q. How can we biochemically assess thiamin status of the body ?

Ans. Thiamin status of the body can be assessed by measuring (*i*) transketolase activity in the erythrocytes which is reduced in thiamine deficiency, (*ii*) blood pyruvic acid concentration which is raised in thiamine deficiency.

Q. What are important sources of thiamin ?

Ans. Rich sources of thiamin include cereal grains, rice-polish, yeast and liver.

Q. What is daily requirement of thiamin ?

Ans. Average daily requirenment of thiamin, for an adult, is 1.2 – 1.8 mg.

Requirement of thiamin, however, depends upon carbohydrate intake. It has been defined as 0.4 to 0.5 mg/1000 Kcal. Requirement is increased during fever, hyperthyroidism, and increased alcohol intake. It is also increased during pregnancy.

Q. Describe the structure of riboflavin.

Ans. Riboflavin has an isoalloxazine ring with a ribitol side chain.

It is also called as lactoflavin, due to its high content in milk.

Q. Name coenzyme forms of riboflavin and some enzymes where these are used as a coenzyme.

Ans. Riboflavin is a component of two coenzymes, *i.e.* flavin mononucleotide (FMN) and flavin adenine dinucleotide (FAD) : The later is a dinucleotide of FMN and AMP.

These coenzymes are used in oxidation-reduction reactions and form an integral part of the flavoprotein containing oxidoreductases.

The coenzyme accepts two hydrogen atoms from the substrate which, in turn, gets oxidized with the reduction of the coenzyme.

$$AH_2 + FMN/FAD \longrightarrow FMN.H_2/FAD.H_2 + A$$

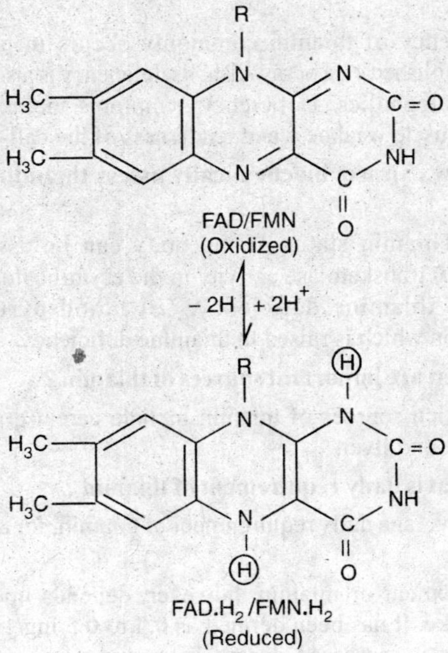

R
N N C = O
H_3C
H_3C
NH
N C
O

FAD/FMN
(Oxidized)

$- 2H$ $+ 2H^+$

R H
N N C = O
H_3C
H_3C
NH
N C
H O

$FAD.H_2/FMN.H_2$
(Reduced)

FMN is used as a coenzyme with cytochrome C reductase and L-amino acid oxidases.

FAD is used with succinate dehydrogenase, glycerol -3-phosphate dehydrogenase, D-amino acid oxidases, etc.

Q. Describe common deficiency symptoms of riboflavin.

Ans. Common deficiency symptoms of riboflavin include cheilosis (lesions at the angle of mouth), glossitis, and ocular as well as skin changes such as localized dermatitis of the face, photophobia and lacrimation.

Q. List important sources of riboflavin.

Ans. Milk, liver and kidney are rich sources of riboflavin.

Q. What is daily requirement of riboflavin ?

Ans. Average daily requirement of riboflavin, for an adult, is 1-2mg or 0.5-0.6 mg/1000 Kcal.

Q. What is niacin ?

Ans. Niacin is a generic name given to two compounds which have a pyridine ring. These are referred to as nicotinic acid and nicotinamide. Both of these substances have niacin activity and are known as pellagra-preventive factors or antipellagra factors.

Q. What are coenzyme forms of niacin ?

Ans. Niacin exists in the body in the form of two coenzymes, *i.e.*—

1. Nicotinamide adenine dinucleotide (NAD), and
2. Nicotinamide adenine dinucleotidephosphate (NADP).

Q. What is biochemical role of coenzymes of niacin?

Ans. Two coenzymes of niacin, *i.e.* NAD^+ and $NADP^+$ exist

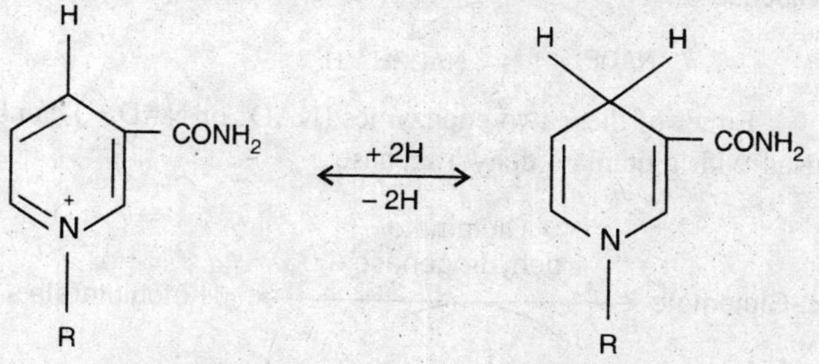

H CONH$_2$ N$^+$ R — +2H / −2H ⇌ — H H CONH$_2$ N R

$NAD^+(P)$ (oxidized) $NAD(P)H + H^+$ (Reduced)

in the oxidised state and carry a positive charge. They accept two H^+, one of which is accepted by the pyridine ring while the other one remains in the solution.

These are used as coenzymes with various dehydrogenases. For example, NAD^+ is used in the conversion of pyruvate to acetyl CoA, α-ketoglutarate to succinyl CoA or malate to oxaloacetate.

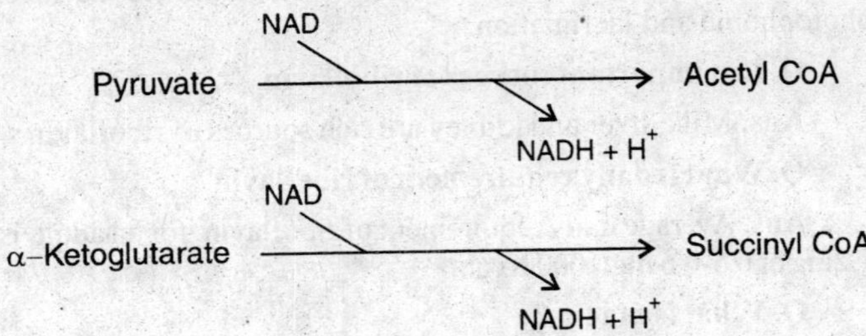

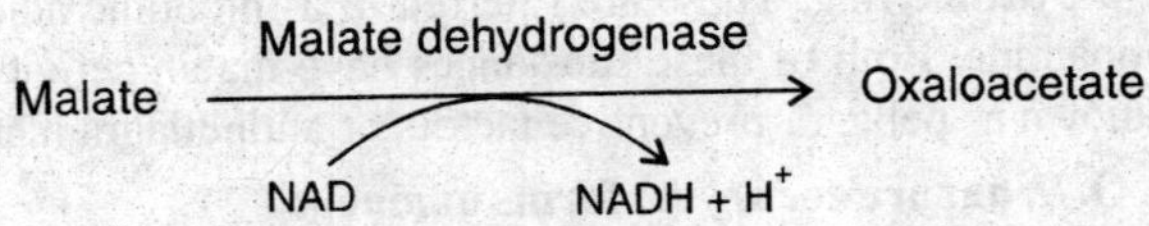

$NADP^+$ is used as a coenzyme in the conversion of glucose–6– phosphate to 6–phosphogluconolactone, by glucose–6–phosphate dehydrogenase in the HMP shunt.

Glucose-6-P dehydrogenase

Glucose-6-P ⟶ 6-Phosphogluconolactone

$NADP^+$ ⟶ $NADPH + H^+$

Either of these two coenzymes (NAD^+ or $NADP^+$) can be used with glutamate dehydrogenase.

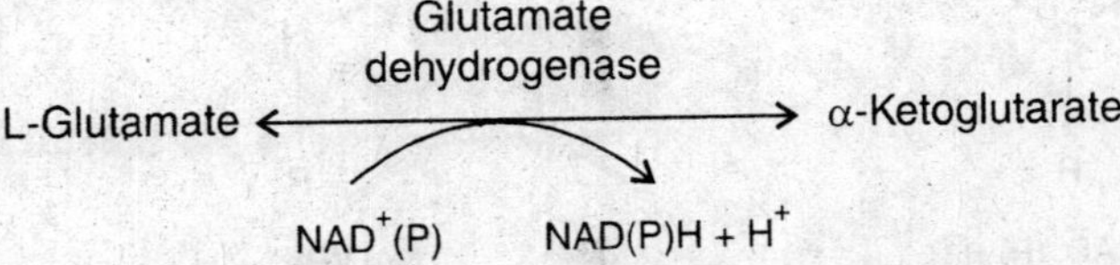

Q. What is pellagra ?

Ans. Pellagra is a disease due to the deficiency of niacin. It is characterised by 3D's, *i.e.* dermatitis, diarrhea and dementia. If not treated, prolonged deficiency of niacin may also result in 4th D, *i.e.* death.

Pellegra is common in people who are consuming diet rich in maize or sorghum as these staple-foods are deficient in tryptophan (a precursor of niacin).

Q. List important sources of niacin.

Ans. Rice polish, yeast and liver are good sources of niacin.

Milk and milk products are good sources of tryptophan which can be converted to niacin in the body.

Q. What is daily requirement of niacin ?

Ans. Average daily requirement of niacin for an adult is 12-20mg. This, however, depends upon the intake of tryptophan .

Q. What is tryptophan-niacin conversion ratio ?

Ans. Sixty mg of tryptophan can produce 1 mg of niacin. Hence, tryptophon-niacin conversion ration is 60 : 1.

Conversion of tryptophan to niacin, however, depends upon pyridoxine status of the body since it is required in the conversion of tryptophan to niacin.

Q. What is pyridoxine ?

Ans. Pyridoxine is also called vitamin B_6.

Three pyridine derivatives, *i.e.* pyridoxine, pyridoxal and pyridoxamine exhibit vitamin B_6 activity.

Q. What is the coenzyme form of vitamin B_6?

Ans. The coenzyme form of vitamin B_6 is called as pyridoxal-5-phosphate.

Q. Describe biochemical role of vitamin B_6.

Ans. Pyridoxal-5-phosphate, the coenzyme form of vitamin B_6 , is mainly used in transamination reactions, *i.e.* with transaminases such as glutamate pyruvate transaminase (GPT) and glutamate oxaloacetate transaminase (GOT).

Pyridoxal-5-phosphate is also used as a coenzyme with dehydratases, desulfhydrases and decarboxylases.

Q. What are deficiency symptoms of vitamin B_6 ?

Ans. Common deficiency symptoms of vitamin B_6 include cheilosis, glossitis and hypochromic anemia.

Its deficiency in infants has been shown to cause demyelination of the peripheral nerves.

Vitamin B_6 is also used in the treatment of nausea and vomiting during pregnancy, radiation therapy and muscular dystrophy.

Deficiency is commonly observed in patients taking isoniazid (INH) during antitubercular therapy.

Q. List important sources of vitamin B_6.

Ans. Egg, fish, green- leafy vegetables and cereals are good sources of vitamin B_6.

Q. What is the daily requirement of vitamin B_6 ?

Ans. Average daily requirement for an adult is about 1-2 mg.

Requirement, however, depends upon protein intake, due to its role in amino acid metabolism.

Q. What is pantothenic acid ?

Ans. Pantothenic acid is a component of coenzyme A . It comprises of a molecule of pantoic acid and β-alanine.

Q. Which vitamin is present in coenzyme A ?

Ans. Pantothenic acid is a constituent of coenzyme A.

Q. Describe the biochemical role of pantothenic acid ?

Ans. Pantothenic acid is a component of coenzyme A. It is required in acetylation reactions, such as in the conversion of pyruvate to acetyl CoA or α-ketoglutarate to succinyl CoA.

It is most important as a coenzyme since it is used in the metabolism of carbohydrates, lipids as well as proteins, *e.g.*—

(*i*) in the Krebs cycle (in the conversion of pyruvate to acetyl CoA)

(*ii*) in the activation of fatty acids as well as in fatty acid synthesis (as a component of acyl carrier protein), and

(*iii*) in oxidative decarboxylation of branched chain amino acids.

Q. Describe deficiency symptoms of pantothenic acid ?

Ans. Deficiency symptoms of pantothenic acid include

headache, fatigue, muscle cramps and GIT disturbances. It is also used in the treatment of burning-feet syndrome, healing ulcers and bedsores.

Q. What are important sources of pantothenic acid ?

Ans. Yeast, liver, egg, potato, whole cereals and legumes are good sources of pantothenic acid.

Q. What is daily requirement of pantothenic acid ?

Ans. Average daily requirement of pantothenic acid is 5-10 mg.

Q. Describe biochemical role of biotin as a coenzyme ?

Ans. Biotin, after forming a complex with the apoenzyme (as active biotin), is used in carbon dioxide fixation reactions, by various carboxylases, *e.g.*—

In the conversion of pyuvate to oxaloacetate or acetyl CoA to malonyl CoA

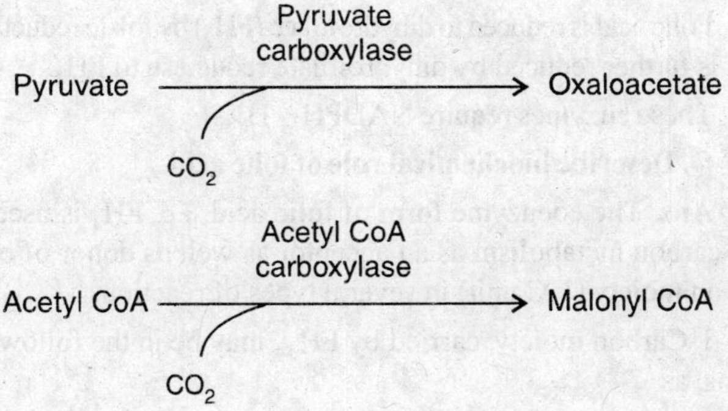

Q. What is the biochemical function of choline in the body ?

Ans. Choline has several important biochemical functions in the body—

(*i*) It is used as a methyl donor for transmethylation reactions,

(*ii*) It has lipotropic activity and prevents fatty liver, and

(*iii*) As acetylcholine, it is a chemical mediator of synaptic transmition.

Q. What is the biochemical function of lipoic acid ?

Ans. Lipoic acid acts as a coenzyme for the transfer of

electrons as well as the activated-acyl group which are released by oxidative decarboxylation of a-keto acids within the multi-enzyme complex. Lipoic acid is first reduced to dihydrolipoic acid which, in turn, acts as an acceptor of the acyl group.

This vitamin is used as a coenzyme with several other coenzymes which include TPP, NAD^+, FAD and CoA, in oxidative decarboxylation, *e.g.* in the conversion of pyruvate to acetyl CoA and α-ketoglutarate to succinyl CoA.

Q. What is folic acid ?

Ans. Folic acid (folate) is a hemopoietic factor. Chemically, it contains pteridine ring, a molecule of para-aminobenzoic acid and several (1, 3 or 7) molecules of glutamate. Hence, folic acid is also called as pteroylglutamic acid.

Q. What is the coenzyme form of folic acid ?

Ans. The coenzyme form of folic acid is called as tetrahydrofolic acid (FH_4).

Folic acid is reduced to dihydrofolate (FH_2) by folate reductase. FH_2 is further reduced by dihydrofolate reductase to FH_4.

These enzymes require NADPH + H^+.

Q. Describe biochemical role of folic acid.

Ans. The coenzyme form of folic acid, *i.e.* FH_4 is used in one-carbon metabolism as an acceptor as well as donor of one-carbon moiety (1-C unit) in several types of reactions

1-Carbon moiety, carried by FH_4, may be in the following forms, as –

Formyl –	–CHO
Formate –	–HCOO
Methenyl –	=CH–
Methylene –	$-CH_2-$
Methyl –	$-CH_3$
Hydroxymethyl –	$-CH_2OH$
Formimino –	–CH=NH

FH_4 may accept one-carbon moiety from histidine (as formimino group), serine (hydroxymethyl group) and choline, thymine or methionine (methyl group).

One-carbon motiety, carried by FH_4, may be used in the synthesis of C_2 and C_8 of purine ring, in the conversion of glycine to serine, uracil to thymine, norepinephrine to epinephrine, etc.

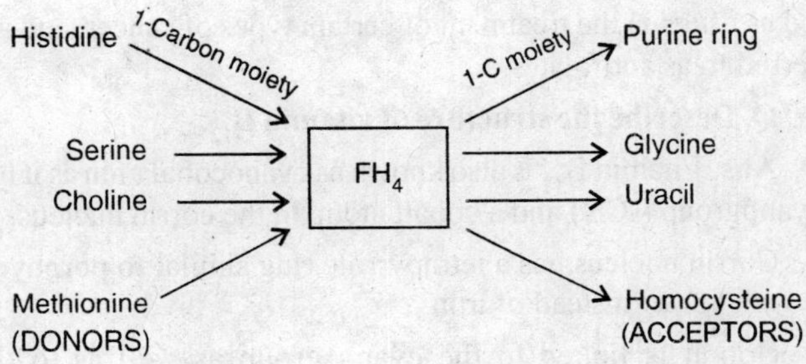

Q. Name deficiency disorders of folic acid.

Ans. Folic acid deficiency may result in macrocytic anemia (megaloblastic anemia) and leukopenia. Deficiency symptoms include—

1. Mental retardation,
2. Growth retardation, and
3. Neural tube defects.

Q. How can we assess folic acid status of the body ?

Ans. Folic acid status of the body can be assessed by measuring urinary exertion of formiminoglutamic acid (FIGLU), after a dose of histidine.

Histidine is catabolized in the liver and forms FIGLU as an intermediate. In the presence of folic acid, FIGLU is converted to N_5–formimino. FH_4 and glutamic acid. But if folic acid is deficient these two compounds are not formed and FIGLU is excreted in the urine in large amounts.

Q. What are important sources of folic acid ?

Ans. Green-leafy vegetables (spinach and cabagge), yeast, liver and kidney are good sources of folic acid.

Q. What is daily requirment of folic acid ?

Ans. Average daily requirement of folic acid, for an adult human being, is 150 to 300 μg.

Q. What are folic acid antagonists.

Ans. Certain analogs of folic acids such as aminopterin and amethopterin are the competitive inhibitors of FH_4 synthesis. Due to their competitive inhibitory effects, these compounds are used as drugs in the treatment of certain types of cancer and are referred to as antifolates.

Q. Describe the structure of vitamin B_{12}.

Ans. Vitamin B_{12} is also known as cyanocobalamin as it has a cyanogroup (-CN) and a cobalt atom, in the corrin nucleus.

Corrin nucleus has a tetrapyrrole ring similar to porphyrin but with cobalt instead of iron.

Cobalt is linked to the cyano group as well as to the imidazole ring.

Q. How is vitamin B_{12} absorbed and metabolized in the body?

Ans. Vitamin B_{12} is absorbed from the gastrointestinal tract in the presence of HCl and a glycoprotein called intrinsic factor. Hence, vitamin B_{12} is also called extrinsic factor.

Vitamin B_{12} is transported in the blood with the protein referred to as transcobalamin. These proteins transport vitamin B_{12} from the liver to various tissues as hydroxycobalamin which can be subsequently converted to methylcobalamin or deoxyadenosyl-cobalamin.

Q. What are the coenzyme forms of vitamin B_{12}?

Ans. Coenzyme forms of vitamin B_{12} are methylcobalamin and deoxyadenosylcobalamin. These are also called cobamides.

Q. What is the biochemical role of vitamin B_{12}?

Ans. Vitamin B_{12} is used in various metabolic reactions.

1. Methylcobalamin is used in the conversion of homocysteine to methionine, whereas

2. Deoxyadenosylcobalamin is used as a coenzyme with methylmalonyl CoA isomerase, in the metabolism of propionyl CoA.

Q. Describe deficiency disorders of vitamin B_{12}?

Ans. Prolong deficiency of vitamin B_{12} may be observed in intestinal malabsorption and atrophy of the gastric mucosa. It results in pernicious anemia and methylmalonic aciduria.

Deficiency symptoms include severe psychosis and extensive mental deterioration.

Q. What are the sources of vitamin B_{12} ?

Ans. Primary source of vitamin B_{12} is its synthesis by the intestinal microbial flora. Besides liver, kidney and fish are good sources of vitamin B_{12}.

Q. What is daily requirement of vitamin B_{12} ?

Ans. Daily requirement of vitamin B_{12}, for an average adult, is 1-3 μg.

Q. Describe biochemical functions of vitamin C ?

Ans. Vitamin C activity is exhibited by ascorbic acid. It can occur in the en-diol form and is used in oxidation-reduction reactions. L-ascorbic acid, in turn, is oxidized to dehydroascorbic acid and is used as a coenzyme in hydroxylation reactions.

Q. Name some reactions where vitamin C is used as a coenzyme.

Ans. Vitamin C is used in several hydroxylation reactions :

(*i*) In the hydroxylation of aromatic amino acids such as :

Phenylalanine ⟶ Tyrosine

Tryptophan ⟶ 5-Hydroxy-tryphophan

(*ii*) In the synthesis of collagen – Vitamin C is used in hydroxylation of proline and lysine in preprocollagen which is converted to collagen.

Proline ⟶ Hydroxyproline

Lysine ⟶ Hydroxylysine

(*iii*) Vitamin C is also essential for the absorption of iron since it helps in the conversion of Fe^{3+} to Fe^{2+}.

(*iv*) It is also important for the synthesis of steroid hormones, due to its role in the hydroxylation of the steroid nucleus.

Q. What is scurvy ?

Ans. Scurvy is a disease due to vitamin C deficiency. It is characterized by :

(*i*) Swollen-spongy and bleeding gums

(*ii*) Multiple hemorrhages

(*iii*) Loosening of teeth and joint pain.

(*iv*) Deficiency of vitamin C may also lead to microcytic anemia.

Q. What are important sources of vitamin C ?

Ans. Amla is a richest source of vitamin C. Citrus fruits, green chillies, guava and tomato are also good sources of vitamin C.

Q. What is daily requirement of vitamin C ?

Ans. Average daily requirement of vitamin C, for an adult human being, is 40 to 60 mg.

Q. Is excess dose of vitamin C harmful for the body ?

Ans. Though vitamin C is a water soluble vitamin, however, if taken in very high doses for a longer duration, it may result in hypervitaminosis C.

It may result in abdominal pain, diarrhea and nausea. It may also result in urolithiasis, due to the formation of oxalate stones since principal end product of its catabolism is oxalic acid.

9

Minerals

Q. What are important biological functions of calcium in the body?

Ans. Calcium performs several functions—

1. Calcification of bone and teeth
2. Clotting of blood
3. Contraction of muscle
4. Activator for some enzymes like adenylate cyclase, ATPase, etc.
5. As a second messenger
6. Permeablity of capillary walls
7. Excitability of nerve fibres.

Q. What are the factors which affect calcium absorption?

Ans. Calcium absorption is affected by several factors—

1. Vitamin D helps in increasing calcium absorption
2. Dietary phosphorus reduces calcium absorption
3. Cereals, due to presence of phytic acid, reduce calcium absorption.
4. Increased amount of dietary fat, crude fibre and oxalate-rich diet reduce calcium absorption.
5. Dietary protein, lactose and acidic food promote calcium absorption.

Q. What is calcium homeostasis?

Ans. Maintenance of serum calcium (Ca^{2+}) levels with in the normal limits is referred to as calcium homeostasis.

More than 50% of calcium is present in serum in the ionised form. Its concentration is regulated in a complex manner by the following factors—

1. **Parathyroid hormone :** PTH stimulates reabsorption of calcium from the kidney tubules. It also increases bone

demineralization and promotes osteoclastic activity of the cell. PTH also stimulates 1-α-hydroxylase in the renal tubules and thus promotes the activation of vitamin D.

The overall effect of all these processes in turn leads to increased serum calcium concentration.

2. **Vitamin D :** PTH stimulates the activation of vitamin D to 1, 25-dihydroxyvitamin D (active vitamin D or calcitriol) which in turn promotes intestinal absorption of calcium by stimulating the synthesis of calcium binding protein.

3. **Calcitonin :** Calcitonin promotes osteoblastic activity in the bones and helps in lowering serum calcium level by promoting mineralisation of the bone.

Both, PTH and active vitamin D thus increase serum calcium level when it is low whereas calcitonin acts in the opposite direction.

Q. What is calmodulin ?

Ans. Calmodulin is a calcium-binding protein present in eukaryotes. It acts as a calcium receptor in the cell and pumps out Ca^{2+}.

Q. Define total body content, normal serum concentration, dietary requirement and sources of calcium ?

Ans. Adult human body contains nearly 1000 to 1200 g of calcium. Mainly, it is present in bones and teeth.

Normal serum calcium level is 9–11 mg %.

Daily requirement of calcium is 800 mg to 1200 mg.

Milk and milk products are rich sources of calcium. Egg yolk, dry fruits and legumes are its good sources.

Q. What is hypocalcemia. List clinical disorders related to hypocalcemia.

Ans. Hypocalcemia refers to decreased Ca^{2+} concentration in serum. It may be a result of low dietary calcium, lack of vitamin D or hypoparathyroidism.

Prolonged hypocalcemia may result in rickets in children or osteomalacia in adults.

Hypocalcemia may also lead to osteoporosis and tetany.

Q. What is osteoporosis ?

Ans. Osteoporosis is characterized by loss of bone matrix, due to progressive demineralization. This in turn may lead to increased fragility and fracture of bones, mainly of the extremity and pelvis, particularly, during menopause and old age.

Increased dietary calcium, dietary nutrients which helps in the formation of bone matrix such as Mg, P, Zn, Cu, vitamin D and proteins along with regular exercise are essential in the prevention of osteoporosis. In postmenopausal woman, estrogen therapy may also be given.

Q. What is tetany ?

Ans. Tetany is a condition characterized by neuromuscular excitability and convulsive seizures. It may be a result of hypocalcemia, hypomagnesemia and alkalosis.

Q. What is hypercalcemia ?

Ans. Hypercalcemia is referred to as increased serum calcium concentration.

It may be a result of increased dietary absorption of calcium, increase in plasma albumin, or increased renal reabsorption of calcium. Hyperparathyroidism and/or hypervitaminosis D may result in hypercalcemia. Increased serum calcium in turn may lead to the formation of calcium stones. It may also cause loss of appetite, constipation and sluggish reflex activity of CNS.

Q. List important biological functions of phosphorus in the body.

Ans. 1. Phosphorus, along with calcium is important for the mineralization of bone and teeth, as it forms a part of the hydroxyappetite crystal.

2. It is important in the maintenance of acid base balance and pH of body fluids.

3. As an important constituent of phospholipid, it is essential for the transport and metabolism of lipids. It also forms integral part of the cell membrane.

4. As a constituent of nucleotides (nucleic acids), phosphorus is also essential for cell multiplication and cellular functions.

5. It acts as an activator in phosphorylation reactions.

6. As a constituent of high energy phosphate bond (ATP), it helps in energy metabolism.

Q. What is average total phosphorus content of body in an adult ? Describe various forms in which phosphorus is present in the body ?

Ans. Normal adult human being has nearly 600–800 g of phosphorus. Nearly 75% of it is present in bones, teeth and skeletal muscle.

Phosphorus is present in the body in two forms—

1. **Inorganic phosphorus :** Which is essential for the formation and development of bones and teeth alongwith calcium. It is also present with sodium and potassium in the soft tissues and body fluids.

2. **Organic phosphorus :** Such as high energy compunds (ATP, creatine phosphate), phospholipids (lecithins), nucleic acids and sugar esters (glucose-6-phosphate), etc.

Q. What are various factors which affect absorption of phosphorus ?

Ans. About 10–40% of dietary phosphorus is absorbed in the small intestine as inorganic phosphate. Absorption of phosphorus depends upon several factors—

1. **Type of food :** Phosphorus is absorbed in a greater amount from animal food compared to cereals and legumes, as phytic acid present in cereals reduces its absorption.

2. **Dietary factors :** Such as calcium, vitamin D and phosphorus content of the diet.

Q. What is normal serum phosphorus concentration and how it is regulated?

Ans. Normal level of serum phosphorus (as inorganic phosphorus) varies from 2.5 to 4.5 mg/100ml.

Parathyroid hormone and serum calcium regulate serum phosphorus concentration. Increased dietary concentration of phosphorus reduces its absorption.

Q. What is daily requirement of phosphorus. List its important sources ?

Ans. Daily requirement of phosphorus for normal adult varies from 1000–1500 mg.

Foods rich in calcium, such as milk and milk products, are also rich sources of phosphorus. It is also found in eggs, wheat and other cereals.

Q. What is hypophosphatemia ?

Ans. Low serum phosphate concentration is referred to as hypophosphatemia.

For normal calcification or mineralisation of bones, calcium to phosphorus product should be 40 (10×4). If the product is below 30 it may lead to rickets/osteomalacia.

Q. What are important biological functions of magnesium ?

Ans. 1. Magnesium, along with calcium and phosphorus, is a component of the bone structure.

2. Magnesium (Mg^{2+}) acts as an activator for several enzymes, generally those requiring ATP such as glucokinase, hexokinase, etc.

3. It plays a role in nerve impulse transmission.

4. Mg^{2+} helps in cardiac and smooth muscle contraction.

Q. Name important iron containing compounds found in the body ?

Ans. Iron is present in the body with heme as well as nonheme containing proteins.

Heme containing proteins include hemoglobin, myoglobin, catalase, peroxidases, etc. Nonheme proteins include ferritin, transferrin, cytochromes, etc.

Q. What are important biological functions of iron ?

Ans. 1. It helps in transport of oxygen with hemoglobin and myoglobin.

2. Iron forms an essential component of metalloenzymes such as catalase, peroxidase, tryptophan pyrrolase, etc.

3. It is required in electron transport chain and oxidative phosphorylation, as a component of cytochromes.

4. Iron, as a component of myeloperoxidase (lysosomal enzyme), is essential for phagocytosis.

5. It is also a component of several nonheme proteins such as ferridoxin, transferrin, etc.

Q. What is Ferridoxin ?

Ans. Ferridoxins are nonheme iron containing proteins, also referred to as iron-sulphur centres, which form a component of the electron transport chain. These proteins contain equal number of iron and sulphide ions. Due to the presence of iron, Fe.S centres exist both in the oxidised as well as reduced state.

Complex-I (NADH dehydrogenase) contains 6 to 7 Fe.S centers.

Q. How is iron absorbed ?

Ans. Iron is present in the food in the ferric form (Fe^{3+}). With the help of HCl of the gastric juice and other reducing substances such as vitamin C, Fe^{3+} is reduced to Fe^{2+}. Fe^{2+} chelates with ascorbic acid, certain sugars and amino acids, which remain soluble in the alkaline fluid of jejunum and duodenum. In the mucosal cells, Fe^{2+} is again oxidised to Fe^{3+} by a copper containing protein referred to as ceruloplasmin (ferroxidase-I). It is transferred to plasma and binds either to apotransferrin or apoferritin. Unabsorbed iron is excreted in the feces.

The mechanism of absorption of iron in the mucosal cell is called mucosal block theory, since absorption of iron is blocked when plasma iron is high and vice-versa. It is the apoferritin, also called as mucosal receptor protein, which controls the absorption of iron. After apoferritin binds with iron, it is converted to ferritin high concentration of which blocks the absorption of iron in the mucosal cells. During iron needs, apoferritin is rapidly synthesized and becomes readily available for the increased absorption of iron.

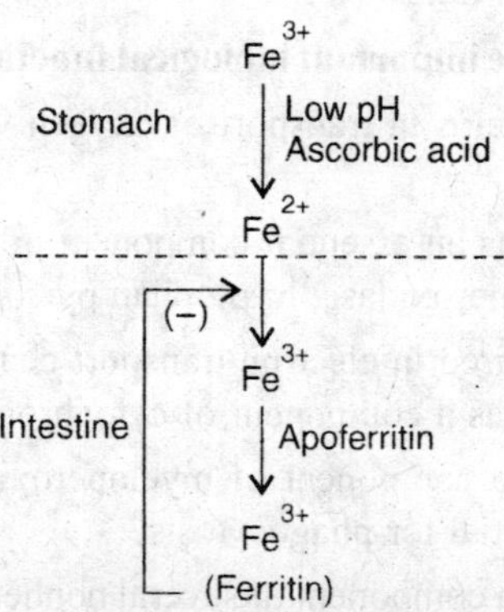

Q. Write note on ferritin ?

Ans. Ferritin is an iron containing protein. It is synthesized in the liver as apoferritin.

Apoferritin is a multimeric protein consisting of 24 subunits of H-chains and L-chains which are arranged spherically and form a polypeptide shell. Its synthesis is regulated by free intracellular iron concentration. Each molecule of apoferritin can take-up large amount of iron (1500 to 2000 atoms of iron).

Ferritin acts as a source of iron during its need by the body. It also acts as a sink for iron during excess intake, *i.e.* when iron is in excess, its storage capacity is increased and apoferritin is rapidly synthesized.

Q. What are ferroxidases ?

Ans. Ferroxidases are a group of copper containing proteins which help in the oxidation and reduction of iron. There are two enzymes referred to as ferroxidase I and ferroxidase II which help in the interconversion of Fe^{2+} to Fe^{3+} and vice-versa, for its dietary absorption and incorporation into different proteins.

Ferroxidase I is identical with ceruloplasmin, a copper containing plasma protein.

Q. What is transferrin ?

Ans. Transferrin is a non-heme iron protein synthesized in the liver as apotransferrin. Apotransferrin is a glycoprotein (β_1-globulin) and has two iron binding sites. It helps in the transport of iron.

Iron is bound to apotransferrin as Fe^{2+} and gets oxidized to Fe^{3+}. In a normal healthy adult, nearly one-third of the iron binding sites are occupied. Plasma transferrin is in equilibrium with hemoglobin in bone marrow and with ferritin in the liver. Whenever there is a need for the synthesis of hemoglobin in the bone marrow, Fe^{3+} is released from transferrin, reduced to Fe^{2+} and gets incorporated into heme. On the other hand, when there is a need of iron by the body such as after hemorrhage, liver and spleen release their iron stores.

Q. What is hemosiderin ?

Ans. Hemosiderin is an iron storage protein present in the form of brownish granules with in the cell. Excess of iron, such as during iron over load, stimulates storage capacity of the newly synthesized apoferritin. This, in turn, gets partially degraded by the proteolytic enzymes of the lysosomes and results in the release of iron oxide micelles, which form insoluble aggregates and get trapped in the secondary lysosomes. These insoluble aggregates of amorphous iron are referred to as hemosiderin.

Excessive deposition of these iron granules in different tissues such as liver and spleen, is referred to as hemochromatosis.

Q. What is daily requirement of iron ? List its important sources.

Ans. Recommended dietary allowance of iron, for a normal adult male, is about 10 mg/day. It is nearly 18 to 20 mg/day for normal adult females during the reproductive stage of their life. This is further increased to 25 to 35 mg/day during pregnancy and lactation.

Foods cooked in iron utensils are the best sources of iron. Green-leafy vegetables, cereals and legumes are also good sources of iron. Milk is a poor source.

Q. What are the deficiency disorders of iron meta-bolism ?

Ans. Most common iron deficiency disorder is anemia. It is a common feature in children, adolescent girls and females during the child-bearing age, particularly, in the developing countries. Such individuals become dull and inactive, loose appetite and have reduced work capacity. It is also associated with decreased immunocompetence. Hookworm infection may also lead to anemia.

Anemia may be of three types—

1. **Microcytic anemia :** It is caused by iron deficiency. Erythrocytes become small in size and have reduced hemoglobin.

2. **Macrocytic anemia :** It is usually caused by the deficiency of hemopoietic factors, *i.e.* vitamin B_{12} and/or folic acid. Size of the RBCs is increased whereas their number is reduced. This is referred to as megaloblastic anemia. Deficiency of vitamin B_{12} along with intrinsic factor, in turn, may lead to pernicious anemia.

3. **Normocytic anemia :** It is usually a result of loss of iron from the body, due to internal/external hemorrhage. Size of the erythrocytes as well as their hemoglobin content remains normal but number of the circulating red blood cells is decreased.

Q. What is hemochromatosis ?

Ans. Hemochromatosis, also referred to as hemosiderosis, is a condition in which excessive amount of iron is deposited in the form of brownish granules. These aggregates of amorphous iron are designated as hemosiderin. Iron gets deposited in this form in many tissues, such as liver and spleen, resulting in hepatomegaly, pancreatic and cardiac dysfunctions, and pigmentation. This may be a result of iron overload, alcoholic cirrhosis, chronic pancreatitis or excessive blood transfusions.

Q. What is the significance of copper for the body ?

Ans. 1. Copper is a component of several enzymes such as cytochrome oxidase, tyrosinase, superoxide dismutase, etc.

2. It is also an important constituent of various tissue specific proteins found in the bone marrow, brain and liver, *i.e.* erythrocuprin, cerebrocuprin and hepatocuprin, respectively.

3. Copper also helps in the formation of bone and elastin fibres, and myelin sheath in the nerve fibres.

4. It is also a constituent of δ-aminolevulinate synthetase and ferroxidase I (ceruloplasmin). Copper, thus helps in the absorption and metabolism of iron.

Q. What is ceruloplasmin ?

Ans. Ceruloplasmin is a copper containing α_2-globulin, found in plasma. It is synthesized in the liver as apoceruloplasmin which can bind six to eight atoms of copper and gets converted into ceruloplasmin.

Ceruloplasmin has ferroxidase activity and helps in the conversion of Fe^{2+} to Fe^{3+}, for its incorporation into the apo-transferrin and apoferritin.

Q. What is Menke's disease or kinke hair syndrome ?

Ans. Menke's disease is an X–linked hereditary disorder, due to failure of copper absorption and transport. As a result, serum copper concentration is reduced. It is characterized by skeletal malformation, mental retardation, defective thermoregulation and immunological disturbances. Most of these abnormalities are due to reduced activities of copper containing oxidases, such as tyrosinase, lysyloxidase, etc.

It is also referred to as kinke-hair syndrome, due to the failure to keratinize hair which become kinke and steel-grey.

Q. What is Wilson's disease or hepatolenticular degeneration ?

Ans. It is an autosomal recessive inherited disorder, due to positive copper balance. The disease is due to a defect in the incorporation of copper into apoceruloplamin, resulting in the deficiency of ceruloplasmin with copper toxicity.

Copper gets accumulated in several tissues such as liver and basal nuclei of the brain. Hence, Wilson disease is also known as hepatolenticular degeneration. Copper toxicity also results in its increased excretion, causing renal tubular damage with hemoglobinuria and generalized amino aciduria.

Q. What is the significance of iodine in the body ?

Ans. Iodine is an important constituent of iodine containing hormones, which are synthesized and released from the thyroid gland.

Q. What are iodine deficiency disorders ?

Ans. Dietary iodine deficiency causes depletion of iodine stores of the thyroid gland thereby reducing the levels of thyroid hormones. This, in turn, stimulates the release of thyroid stimulating hormone (TSH) by the pituitary, causing hyperplasia of the thyroid gland. This is referred to as goitre.

Q. What is myxedema ?

Ans. Severe hypothyroidism, in adults, may result in a condition referred to as myxedema. Such individuals are hypersensitive to cold and mentally retarded. Due to deposition of the myxomatous material in the subcutaneous layer of the face and hands, skin becomes thick and puffy. Mean body weight is also increased due to the retention of water and salt.

Q. What is cretinism ?

Ans. Cretinism is a condition of hypothyroidism in infants and children. Children become dwarf, mentally-retarded and sexually-underdeveloped.

Q. What is thyrotoxicosis ?

Ans. Thyrotoxicosis is a condition due to hyperthyroidism. It is a result of increased intake of iodine by the thyroid gland.

Q. What is dental caries ?

Ans. Dental caries is a result of flouride deficiency. It is characterized as decalcification of tooth enamel, due to acid formation by the micro-organisms.

Breakdown of the tooth enamel, in turn, exposes the dentine and leads to the development of caries.

Q. What is fluorosis ?

Ans. Excess fluoride intake may result in its toxicity, referred to as fluorosis.

Fluoride toxicity may manifest as endemic fluorosis and result in dental or skeletal deformities.

Dental fluorosis occurs, mainly, in children before completion of the mineralization of teeth enamel. Such teeth exhibit mottling which is characterized by minute abnormal white-flacks alongwith yellow or brown spots, scattered all over the tooth surface.

Skeletal fluorosis results in pain, inflammation and restricted movements of the joints along with stiffness of the spine, in adults.

Q. What is genu vulgum ?

Ans. It is the severe form of skeletal fluorosis due to a continuous high intake of fluoride. In addition to dental and skeletal fluorosis, such individuals also show extensive osteoporosis of the limbs, particularly the bones of the lower limbs.

Q. What is the significance of mangenese ?

Ans. Mangenese is a cofactor for several enzymes such as phosphorylase, phosphotransferases, decarboxylases, etc.

Mangenese is also required in the synthesis of glycoproteins in the cartilages.

Q. What is the significance of chromium in the body ?

Ans. Chromium helps in the regulation of blood glucose and acts as a potentiator of insulin secretion.

It has been shown to be a component of a complex referred

to as glucose tolerance factor. It has also been shown to help in the metabolism of lipoproteins.

Q. What is the significance of selenium in the body ?

Ans. Selenium is an essential component of several proteins referred to as selenoproteins.

An important selenoprotein in mammals is glutathione peroxidase which protects the cell against oxidative damage.

Q. What are selenoproteins ?

Ans. Selenoproteins are selenium containing proteins. Selenium is incorporated into peptide chain as selenocysteine (SeC) during protein synthesis. These SeC residues participate in redox reactions and prevent the cell against oxidative damage. Glutathione peroxidase is a selenoprotein found in animals.

Q. What is selenosis ?

Ans. Selenium toxicity is referred to as selenosis. Workers in the paint, glass and ceramic industry inhale excess amount of selenium which, in turn, results in difficult breathing and primary neoplasms of reticuloendothelial system.

Q. What are biological functions of cobalt ?

Ans. 1. Cobalt is an important constituent of cobalamin (vitamin B_{12})

2. Cobalt also increases the production of erythropoietin.

Q. What is the biological significance of molybdenum ?

Ans. Molybdenum is an integral part of several flavoprotein containing enzymes such as xanthine oxidase, sulphide oxidase, etc.

Q. What are important functions of zinc in the body?

Ans. 1. Zinc is a component of several enzymes such as carbonic anhydrase, alcohol dehydrogenase, alkaline phosphatase, RNA polymerase, DNA polymerase and retinene-retinal reductase.

2. Zinc is also important in vision cycle. It stimulates the release of vitamin A from the liver into the blood and thus maintains plasma vitamin A concentration.

3. Zinc is also important for taste sensation, as an important constituent of gustin (a salivary polypeptide).

10

Water and Electrolytes

Q. Describe various physiological functions of water in the body.

Ans. Water has several physiological functions in the body:

1. Water acts as a medium for the transport of nutrients and biomolecules.

2. It acts as a solvent and helps in the elimination of waste products, and metabolites of xenobiotics and drugs.

3. Water provides a hydration shell for the colloidal molecules.

4. It helps in maintenance of the extracellular and intracellular fluids in the body.

5. Water facilitates bowel movement.

6. It provides lubrication in different organs within the body, *e.g.* synovial fluid between the joints.

7. Water helps in the regulation of body temperature.

Q. How is water distributed in different compartments of the body?

Ans. Total body water (35-40 L) is distributed in two compartments, as intracellular fluid and extracellular fluid.

Intracellular fluid : Water present with in the cells is referred to as intracellular fluid. It constitutes about 2/3rd of total body water in adults.

Extracellular fluid : It is the water present extracellularly. It comprises of plasma and interstitial fluid.

Plasma : Water circulating in the blood vessels (intravascular fluid) is referred to as plasma. It is about 5% of the total body volume. Total plasma volume is between 3–4 L in a normal adult.

Interstitial fluid : It is the fluid present outside the blood vessels and comprises of nearly 15% of total body volume.

Interstitial fluid is the ultrafiltrate of plasma and separated from it by endothelium lining of the capillary which acts as a semi-permeable membrane.

Q. What is Gibbs-Donnan equilibrium ?

Ans. Gibbs-Donnan equilibrium is the phenomenon that explains the process of the establishment of an equilibrium between two solutions which are separated from each other by a semipermeable membrane.

When two solutions are separated from each other by a semipermeable membrane and their solutes are freely movable across the membrane, various ions get equally distributed in two compartments. This, in turn, results in the state of equilibrium at which total concentration of the osmotically active particles become equal on two sides of the membrane. If solution on one side of the membrane contains ions which are impermeable to the membrane, *e.g.* proteins, distribution of the diffusible ions at the steady state in equilibrium will be unequal but the ionic product of the concentration of diffusible ions on one side will be equal to the product of the diffusible ions on the other side.

Q. List important dietary sources of water.

Ans. Important dietary sources of water include—

(*i*) Beverages, fruits and vegetables, and

(*ii*) Its metabolic production.

Oxidation of dietary nutrients results in the production of water with in the body, as follows—

Dietary nutrient	Amount of water produced (ml/100g)
Fat	107
Carbohydrate	56
Protein	41

On an average, nearly 300ml of water is produced endogenously, every day.

Q. What is daily requirement of water ?

Ans. Average daily requirement of water, for a normal adult, at moderate environmental temperature and humidity is nearly 2500 ml/day, *i.e.* about 1ml/Kcal.

A healthy infant requires higher amount of water, *i.e.* about 1.5 ml/Kcal.

Daily need, however, varies widely with respect to body surface area, body fat, metabolic rate, physical activity and environmental factors like temperature, humidity, etc.

Q. How is water content of the body regulated ?

Ans. In a normal adult, the quantity of water intake is balanced by the quantity which is eliminated from the body. Thus water balance is regulated by intake as well as excretion.

Water intake, in turn, is regulated by thirst which is under the control of thirst centre. It, however, also depends upon several other factors, such as obligatory losses, pathological conditions and dietary factors.

Obligatory losses such as insensible perspiration, breath and feces affect fluid loss, hence water intake.

Fluid intake is also associated with physical conditions and injury, *e.g.* increased fluid loss due to diarrhea or blood loss, in turn, increases water intake.

Water balance is also affected by various dietary constituents, *e.g.* intake of sodium, carbohydrates and proteins.

Water output is affected by several factors such as volume of urine, feces, and sensible as well as insensible perspiration.

Q. Describe various mechanisms which regulate electrolyte balance.

Ans. Various mechanisms which maintain electrolyte balance include Na^+, K^+– pump, kidney and environmental factors.

1. **Na^+, K^+ – Pump :** This maintains low Na^+ and high K^+ concentration, intracellularly.

2. **Kidney :** Kidneys regulate Na^+ as well as water excretion, depending upon dietary intake as well as physiological conditions.

3. **Environmental conditions :** Conditions such as temperature, exercise, etc., regulate thirst as well as electrolyte and water balance.

Q. What are important functions of sodium in the body?

Ans. Important functions of sodium in the body include—

1. Regulation of osmotic pressure.
2. Regulation of acid-base balance.
3. Absorption of monosaccharides and amino acids.
4. Nerve transmission.
5. Maintenance of blood viscocity.
6. Maintenance of water and electrolyte balance
7. Na^+ also acts as a cofactor for Na^+, K^+ – ATPase.

Q. What is Na^+, K^+–pump ?

Ans. Na^+, K^+ – pump (Na^+, K^+– ATPase) transports Na^+ from the intracellular compartment to the extracellular compartment against electrochemical gradient. It also transports K^+ intracellularly. The pump requires energy which is obtained from the hydrolysis of ATP. This in turn helps in maintaining high concentration of sodium in the extracellular fluid and potassium in the intracellular fluid.

Q. Explain processes which regulate sodium concentration.

Ans. Sodium concentration is regulated by several factors, such as—

1. **Hormones :** Aldosterone facilitates sodium retention in the body.

2. **Disorders of the GIT :** Vomitting and diarrhea results in loss of sodium along with water.

3. **Temperature :** In a hot, temperate climate large amount of sodium is lost in perspiration.

4. **Renal functions :** In renal tubular damage, urinary excretion of sodium is increased.

Q. What is daily requirment of sodium ?

Ans. Daily intake of sodium (as sodium chloride), varies from 5 to 15 g and depends upon climatic conditions.

Q. List important sources of sodium.

Ans. Most important source of sodium in the diet is common salt (NaCl). Besides it, meat, seafood, processed cheese and bread are good sources of salt.

Q. What is hyponatremia ?

Ans. Hyponatremia is defined as a decrease in plasma Na^+ concentration below 135 meq/L (135 mmol/L).

Hyponatremia may lead to loss of appetite, and muscular cramps of the extremities and the abdomen. Hyponatremia accompanied with the loss of extracellular fluid occurs through renal or extra renal losses.

Renal losses can occur with osmotic diuresis, adrenal insufficiency or potassium sparing diuretics.

Q. What is Hypernatremia ?

Ans. Hypernatremia is defined as an increase in plasma Na^+ concentration above 150 meq/L (150 mmol/L). It is always hyperosmolar and is characterized by edema, anemia and azotemia. It may also result in essential hypertension.

Q. List important physiological functions of potassium.

Ans. 1. Potassium has an important role in the regulation of blood pH, as potassium salt forms an important buffer system in the intracellular compartment.

2. It also helps in the maintenance of osmotic pressure and electrolyte balance.

3. It also helps in the maintanence of neuromuscular irritability and excitablity.

4. It forms an essential component of the digestive juices such as pancreatic juice and bile.

5. It acts as a cofactor for some of the enzymes, such as pyruvate kinase.

Q. How is potassium excreted and regulated ?

Ans. Potassium is excreted almost entirely by the kidneys.

Several factors which affect the excretion of potassium include—

1. Its cellular content,

2. GIT disturbances, and

3. Renal functions.

Q. What is daily requirement of potassium ? Also list its important sources.

Ans. About 4 g of potassium (as potassium chloride) is required daily.

Vegetables, dried beans, coffee and coco are good sources of potassium.

Q. Describe hyperkalemia.

Ans. Hyperkalemia is characterized by serum potassium level above 5.5 meq/L.

Hyperkalemia may lead to—

1. CNS depression,
2. Mental confusion,
3. Numbness of the extremities,
4. Weakness of respiratory muscles, and
5. Bradycardia.

Q. Describe hypokalemia.

Ans. Hypokalemia is defined as serum potassium level below 3.5 meq/L.

Reduced level of plasma potassium may result in arrhythmia, tachycardia and hypotension. Potassium deficiency may also result in muscular weakness, bone fragility, paralysis, decreased growth rate and loss of body weight.

Q. List important physiological functions of chloride.

Ans. 1. Chloride, along with sodium and potassium, helps in maintaining normal osmotic pressure of body fluids.

2. It is a major anion and thus participates in maintaining pH of body fluids.

3. It is an essential component of the gastric HCl.

4. It acts as an activator for enzymes like salivary amylase.

Q. What is chloride shift ?

Ans. Electroneutrality of plasma which is caused by the shifting of HCO_3^- from the red blood cells is balanced by the movement of Cl^- from Plasma to RBC's. This process of shifting of chloride into the RBC's, in exchange with HCO_3^-, is called chloride shift.

Q. Define hypochloremia.

Ans. Hypochloremia refers to decreased concentration of

chloride in plasma. It is generally observed during diarrhea, excessive sweating, salt-loosing nephritis, metabolic acidosis and prolonged vomitting. Hypochloremia is also seen in respiratory alkalosis due to hyperventilation.

Q. What is hyperchloremia?

Ans. Increase in plasma Cl^- concentration is referred to as hyperchloremia.

It is observed in patients with—

1. Dehydration
2. Renal tubular acidosis
3. Acute renal failure
4. Diabetes insipidus
5. Cushing's syndrome.

Q. What is edema ?

Ans. Edema is a condition referred to as fluid retention. Due to the retention of water along with sodium, both water as well as sodium move to the interstitial fluid for the maintanence of blood volume.

Retention of sodium and water may be observed in conditions such as congestive heart failure. It may be characterized by puffiness of fingers, ankles or entire body, pulmonary edema and respiratory distress.

11

Nutrition

Q. Name proximal principles of food.

Ans. Proximal principles of food include carbohydrates, lipids, proteins, vitamins, minerals and water.

Carbohydrates, lipids and proteins are the sources of energy. Vitamins and minerals are essential for various metabolic processes. Water is necessary for providing aqueous medium.

Q. What is the unit of energy ?

Ans. The unit of energy is calorie. Energy stored in various food-stuffs is converted to heat and is expressed in terms of calories.

Q. What is a calorie ?

Ans. A calorie is defined as the amount of heat which is required to raise the temperature of 1 g of water by 1°C, *i.e.* from 14.5°C to 15.5°C.

1 Calorie = 4.18 Joule

Q. What is the international unit of energy ?

Ans. International unit of energy is joule.

A joule is defined as the amount of heat that is required to move a mass of 1 kilogram by 1 meter distance, by a force of 1 newton.

Q. What is calorimetery ?

Ans. Calorimetery is a technique of the measurement of energy expenditure. It may be of two types, *i.e.* –

1. Direct calorimetery, and
2. Indirect calorimetery.

Direct calorimetery : It is the method by which amount of heat produced is measured directly, by recording the amount of heat that is transferred to circulating water, around an insulating chamber.

Indirect calorimetery : It is the assessment of the quantity of heat produced by the body, as a result of the combustion of food with reference to the quantity of oxygen consumed.

Q. What is respiratory quotient ?

Ans. Respiratory quotient (RQ) is the ratio of the volume of CO_2 expired to the volume of O_2 utilized.

$$RQ = \frac{CO_2 \text{ expired}}{O_2 \text{ utilized}}$$

RQ for different foodstuffs varies with respect to their oxygen contents. Carbohydrates have highest RQ since the volume of CO_2 eliminated is equal to the volume of O_2 consumed.

$$C_6H_{12}O_6 + 6O_2 \rightarrow 6H_2O + 6CO_2$$

R.Q. of various foodstuffs is as follows :

Foodstuff	R.Q.
Carbohydrates	1.0
Lipids	0.7
Proteins	0.8
Mixed food	0.82

Q. Define calorific value of different food-stuffs ?

Ans. Calorific value of food is defined as the amount of heat (energy) produced on complete combustion of 1 g of food.

Calorific value of different food-stuffs is :

Foodstuff	Calorific value (Kcal/g)
Carbohydrates	4.0
Lipids	9.0
Proteins	4.0
Alcohol	7.0
Wheat	3.5

Q. What are the factors which affect energy requirement ?

Ans. Energy requirement depends upon—

1. Body surface area : Larger the surface area, greater is the amount of heat is loss.

2. Age and sex : Age and sex of an individual affect basal metabolic rate (B.M.R.)

3. Physical Activity : The level of physical activity affects the amount of energy expenditure.

Q. What is basal requirement of energy ?

Ans. Energy requirement at complete physical and mental rest is referred to as basal requirement. Basal requirement of energy is expressed as BMR (basal metabolic rate).

Q. What is BMR ?

Ans. BMR is defined as the rate of energy utilization in a resting state. It is determined experimentally, when a subject is lying down at complete physical and mental rest (with closed-eyes but awaken), wearing light clothes, and at a room temperature of 20–25°C. BMR is expressed as Kcal/m^2/hr.

Average BMR of a normal adult male is 40 Kcal/m^2/hr and for a female is 36 Kcal/m^2/hr.

Q. List various factors affecting BMR ?

Ans. Several factors which affect BMR, include —

(*i*) Age : BMR increases with age upto 5 yrs and there after again at puberty. Nearly 2% reduction in BMR is observed per decade of life, after puberty.

(*ii*) Sex : Women have low BMR than men

(*iii*) Climate : BMR is high in cold than temperate climate.

(*iv*) Physiological state of the body : BMR increases during stress and trauma. It remains unaltered during early pregnancy but increases after first-half of pregnancy, due to the additional BMR of the fetus.

(*v*) Diet : Vegetarians have low BMR. Caffiene, nicotine and alcohol increase BMR. It is reduced during starvation.

(*vi*) Hormones : Thyroxine, sex hormones, growth hormone, epinephrine and corticoids increase BMR.

(*vii*) Clinical disorders : BMR increases during hypermetabolic disorders, *e.g.* fever, leukemia, polycythemia, etc.

Q. What is specific dynamic action (SDA) ?

Ans. Specific dynamic action, also called as thermogenic effect of food, is the amount of heat produced, over and above its calorific value. SDA value for different food-stuffs is as follows –

Food stuff	S.D.A. (%)
Protein	30
Fat	12
Carbohydrates	5
Mixed Food	10

Protein has maximum SDA (30%), due to the reason that digestion and absorption of the protein requires extra amount of energy for various metabolic activities such as deamination and urea synthesis.

For calculating energy requirement of an individual, 10% of the total calories intake (basal requirement) is added towards SDA.

Exact cause of SDA though is uncertain, however it accounts for various activities which are performed by the body for the digestion and absorption of nutrients. This is actually the amount of energy spent by the body to extract calories from the food which are released thereafter.

Q. Give an account of energy requirement with respect to physical activity of an individual ?

Ans. Different types of physical activities require different amounts of energy, *e.g.*—

An individual uses 35–40 Kcal/hr while reading, sitting or standing. On the other hand, requirement for walking upstairs is 1000–1100 Kcal/hr. Thus according to the nature of physical activity performed, individuals are divided into three groups—

1. **Sedentary workers :** These are individuals who perform little physical activity, *e.g.* shopkeepers.

2. **Moderate category of workers :** These are the individuals who perform both physical as well as mental activities, *e.g.* students.

3. **Heavy workers:** These are the individuals who perform more of physical activity such as rickshaw-pullers, athletes, etc.

As the work capacity is also affected by the sex, females performing the same type of work require less calories as compared to males.

Q. What is the daily calories requirement of a moderate worker ?

Ans. Daily calories requirement includes BMR, SDA and the requirement for physical activity. Accordingly, daily requirement of a moderate worker, such as a student, is as follows—

Daily calories requirement (Kcal/day)

	Basal	**SDA**	**For physical activity**	**Total (Approx.)**
Males	1500	150	1200	2800
Females	1300	130	800	2200

Q. What is a balanced diet ?

Ans. A balanced diet is defined as a diet, which provides quantitatively as well as qualitatively adequate amounts of all the nutrients for maintaining a good health and regular physical activities.

In a balanced diet, about 65 to 75% of calories should be available from carbohydrates, 15–20% from lipids and 10–15% from proteins.

Dietary carbohydrates should have digestible carbohydrates including polysaccharides (such as starch and glycogen) as well as disaccharides (lactose and sucrose) which are the readily available sources of energy, and the undigestible carbohydrates which provide bulk to the food such as crude fibre (cellulose, lignin, etc.).

Dietry lipids should include triacylglycerols, cholesterol as well as phospholipids. Lipids not only improve palatability of the food but also provide fat soluble vitamins and essential fatty acids.

Dietary protein should include proteins with high biological value which can meet the requirement of essential amino acids. Proteins are not only required for wear and tear or as a source of energy but are also needed for growth and during recovery from prolonged illness.

Daily requirement of protein for infants and young children is higher (2g/kg) than adults (0.8 to 1 g/kg body weight per day).

Q. What are essential fatty acids ?

Ans. Essential fatty acids are polyunsaturated fatty acids which can not be synthesized by the body. Generally, these include—

(*i*) Linoleic acid

(*ii*) Linolenic acid

(*iii*) Arachidonic acid.

They are essential as they can not be synthesized in the body. They are important in the esterification of cholesterol and help in lowering serum cholesterol level.

Arachidonic acid is also a precursor of prostaglandins, leukotrienes and thromboxanes.

Q. What is crude fibre ?

Ans. Crude fibre, also called as dietry fibre or roughage, include undigestible carbohydrates which do not provide energy but form an essential part of the diet for optimum health.

Insoluble dietary fibre such as cellulose, retain water during passage of food along the gut and results in the production of bulky and soft feces which, in turn, avoids constipation and reduces the risk of colon cancer.

Soluble dietary fibre such as gum and pectin are found in legumes and fruits. They bind dietary cholesterol and bile acids, and lower serum cholesterol level. These are also beneficial for diabetics since they delay gastric emptying and lower postprandial rise in blood glucose.

Q. Write a note on nitrogen balance ?

Ans. Nitrogen balance is a comparison between the intake of nitrogen as protein diet and its excretion through various routes such as undigested protein in feces as well as urea and ammonia in the urine.

When daily nitrogen intake is equal to its excretion, the subject is said to be in a zero nitrogen balance.

An individual is in negative nitrogen balance if protein

intake is less than its loss from the body. This, in turn, results in loss of tissue proteins, such as in protein energy malnutrition, high grade fever, chronic infection, etc.

When dietary protein intake is higher than its loss from the body, the subject is said to be in a positive nitrogen balance. This, in turn, results in the storage of protein in tissues,. *e.g.* in a growing child or during pregnancy and lactation.

Q. What is protein energy malnutrition ?

Ans. Protein energy malnutrition is a condition comprising of two different nutritional disorders, *i.e.,* kwashiorkor and marasmus.

Kwashiorkor : Kwashiorkor is a condition arising due to low intake of protein. It usually occurs during postweaning period when a child gets a diet which is adequate in calories but low in protein. Clinical features of kwashiorkor include muscle wasting, edema, anemia, dry and brittle hair, diarrhea and retarded growth.

Marasmus : Marasmus is due to lack of complete food, *i.e.* when a diet is deficient not only in protein but also in calories as well as other nutrients. A marasmic infant is thin, has wasted appearance and is short for age.

Marasmus differs from kwashiokor with respect to the loss of subcutaneous fat.

Though kwashiorkor and marasmus are two different entities of protein energy malnutrition but generally a patient has overlapping features.

Q. What are essential amino acids ?

Ans. Amino acids, which can not be synthesized in the body and have to be included in the diet, are called essential amino acids. These include valine, isoleucine, leucine, lysine, methio-nine, threonine, tryptophan and phenylalanine.

Besides, there are two amino acids, *i.e.* arginine and histidine, which are not essential for a normal adult but become essential during growth, pregnancy and lactation when their demand is increased. These two amino acids are thus called semiessential amino acids.

Q. What is the biological value of a protein?

Ans. Biological value of a protein is the percentage of the protein ingested to the nitrogen retained by the body. It depends upon the amount of essential amino acids, in the diet. Generally, proteins from animal sources have high biological value as compared to those obtained from plant sources.

Q. What is supplementary action of protein ?

Ans. Proteins from plant sources do not contain all the essential amino acids and are limiting in one or more of them. Such a protein may have an excess of the other amino acid, *e.g.* wheat is limiting in lysine but has excess of methionine. On the other hand, pulses lack methionine but are rich in lysine. When a combination of protein from these two sources is taken together, deficiency of one amino acid can be supplemented from the other source. Such action of protein is referred to as supplementary action of protein.

Q. Name some common nutritional disorders ?

Ans.

Nutritional disorder	Causative factor
Kwashiorkor	Protein
Rickets	Vitamin D
Night blindness/ Xerophthalmia	Vitamin A
Beri–beri	Thiamin
Pellagra	Niacin
Scurvy	Vitamin C
Anemia	Iron
Fluorosis	Fluoride

Q. Describe metabolic changes during starvation ?

Ans. Starvation may be referred to as a situation arising due to the nonavailability of food, desire to loose weight by an individual or one's inability to consume food due to certain clinical condition such as trauma or surgery. Starvation results in loss of body weight as well as fat and protein from several tissues. It may also be accompanied with ketosis.

Metabolic changes during starvation vary with the duration of starvation.

1. **In the early stage of starvation,** *i.e.* 4-5 hrs following

meal, as tissues take up glucose, there is a reduction in blood glucose concentration. This, in turn, stimulates release of glycogen and promotes glycogenolysis in the liver. A fall in insulin also causes release of free fatty acids from the adipose tissue. Hence, free fatty acids concentration as well as their oxidation is increased in the liver.

2. **In the postabsorptive phase,** *e.g.,* after an overnight fast, gluconeogenesis is promoted by utilizing amino acids which are released from the breakdown of proteins, mainly in the liver and muscle.

3. **During prolonged starvation,** *i.e.* after 48 hrs, most of the liver glycogen is diminished. Adaptive changes take place. Lipolysis is increased. Increased release of free fatty acids and their oxidation promote the synthesis of ketone bodies.

Several tissues including brain start utilizing ketone bodies, as a source of energy.

Increased oxidation of ketone bodies also reduces plasma bicarbonate concentration, leading to metabolic acidosis. Extensive loss of tissue protein also leads to decreased production and excretion of urea, and other nonprotein nitrogenous substances from the kidney. This, in turn, results in reduced urine volume. Retention of water also causes edema and retention of sodium.

Metabolic acidosis, in turn, also stimulates excretion of H^+ ions, hence there is an increase in NH_4^+ ions.

Q. What is anorexia nervosa ?

Ans. Anorexia nervosa is a form of starvation, characterized by disrupted eating behaviour. It is a mental attitude where an individual does not eat adequate amount of food, may be due to over reaction to mild obesity, fashionable dieting or early menarche. It is a life-threatening psychiatric illness that is observed mainly in adolescent girls. It is difficult to treat, if diagnosis is delayed.

Some of the individuals may also develop bulimia nervosa, a condition where individuals vomit most of the food soon after ingestion, as an alternative to restricted dietary intake.

12

Nucleic Acids and Molecular Biology

Q. What are nucleic acids ?

Ans. Nucleic acids are the macromolecules which are found in a cell, mainly within the nucleus. These are strongly acidic in nature and occur in the cell in association with the basic proteins referred to as histones. They store genetic information and transfer it from one generation to another and/or regulate protein biosynthesis.

Q. What are the constituents of nucleic acids ?

Ans. Nucleic acids are constituted by the repeated monomeric units called nucleotides. Nucleotides for nucleic acids are like monosaccharides for a polysaccharide or amino acids for a protein.

Two types of nucleic acids are found in a cell. These are referred to as :

(*i*) Deoxyribonucleic acid (DNA), and

(*ii*) Ribonucleic acid (RNA).

Q. Name pentose sugars that are found in nucleotides ?

Ans. Depending upon the type of the nucleic acid, two types of pentose sugars are found in nucleotides, *i.e.* —

(*i*) **Ribose**-found in RNA, and

(*ii*) **Deoxyribose**-found in DNA.

Deoxyribose is also called as 2-deoxyribose since instead of an – OH group at C2 in ribose, a molecule of deoxyribose contains only hydrogen, *i.e.*, the oxygen is missing from the carbon atom number two.

Q. Name different nitrogenous bases that are found in nucleotides?

Ans. Nitrogenous bases found in nucleotides are derivatives of either purine or pyrimidine.

A purine ring is a nine-membered ring containing 5 carbon and 4 nitrogen atoms while pyrimidine is six-membered ring with four carbon and two nitrogen atoms.

Purine bases are :

(*i*) Adenine (6-aminopurine), and

(*ii*) Guanine (2-amino-6-oxypurine)

Pyrimidine bases are :

(*i*) Cytosine (2-oxy-4-aminopyrimidine),

(*ii*) Uracil (2, 4-dioxypyrimidine), and

(*iii*) Thymine (2, 4-dioxy-5-methylpyrimidine (5-methyl-uracil)

Out of these three pyrimidines, only two are found in a nucleic acid.

Cytosine and uracil are found in RNA while cytosine and thymine are found in DNA.

Q. What is a nucleoside ?

Ans. A nucleoside is a combination of a nitrogenous base with the sugar.

Generally, N–l of the pyrimidine ring or N–9 of the purine ring is linked to C–1 of the sugar.

Nucleosides containing ribose are called ribonucleosides while those containing deoxyribose are called as deoxyribonucleo-sides, *e.g.* : adenosine, cytidine, deoxyadenosine, deoxycyti-dine, etc.

Q. What is a nucleotide ?

Ans. A nucleotide consists of three constituents, *i.e.* :

(*a*) a nitrogenous base,

(*b*) pentose sugar, and

(*c*) phosphoric acid.

A nucleotide can be said as a phosphoric acid derivative of the nucleoside and may be called as nucleosidemonophosphate.

Phosphoric acid is generally linked to C–5 of the sugar molecule.

Hence, the pentose sugar is sand-witched between the nitrogenous base and phosphoric acid. It is linked through C–l to a nitrogenous base and through C–5 to phosphoric acid, *e.g.* : adenosine monophosphate is a nucleotide with adenine, ribose and phosphoric acid.

Q. Describe biological significance of various derivatives of nucleotides ?

Ans. Nucleotide derivatives perform several biological funcions–

(*i*) Triphosphate derivatives of nucleosides serves as high energy compounds and are used in the biosynthesis of nucleic acids.

(*ii*) Derivatives of nucleotides also function as coenzymes, *e.g.* NAD.

(*iii*) Monosaccharide derivatives of nucleotides are important in metabolic reactions in the body, *e.g.* UDP-glucose.

(*iv*) Cyclic derivatives of nucleotides, *e.g.* cAMP is an intracellular signal transducer and acts as a second messenger.

(*v*) Synthetic analogs of the nitrogenous bases as well as nucleotides have therapeutic significance, *e.g.* 5-fluorouracil, allopurinol, etc.

Q. How various nucleotides are linked in nucleic acids ?

Ans. Various nucleotides are linked together by 3′, 5′-phosphodiester linkages where a hydroxyl group of the sugar molecule of one nucleotide is linked to the hydroxyl group of the sugar of the other nucleotide. As a result of this, one end of the nucleotide has a free hydroxyl or phosphate group (called as the 5′- terminal –OH or phosphate) while the other end of the polynucleotide has an – OH or phosphate group at the 3′ terminus.

Q. How do we differentiate between the position of different carbons of the pentose sugar from that of the base ?

Ans. To differentiate between the position of different carbon atoms of the pentose from that of the base, the carbon atom of the sugar is represented with a mark called as prime (′) where as the carbon atom of the nitrogenous base is referred without prime, *e.g.* C–1 represents first carbon atom of the nitrogenous base while C–1′ denotes the first carbon atom of the sugar.

Q. Describe various features of the Watson-Crick model of DNA.

Ans. Watson-Crick model of DNA has following features :

(*i*) DNA is a polynucleotide containing *d*AMP, *d*CMP, *d*GMP and TMP, and that the Genetic information resides in the sequence, in which these nucleotides are arranged.

(*ii*) DNA is a double helix. It has two strands which are linked together in the opposite directions, *i.e.* two strands run anti-parallel to each other.

(*iii*) Two strands are wound around a common axis, forming a right-handed helix.

(*iv*) Both the strands are held together by hydrogen bonds which occur between the purine of one strand and the pyrimidine of the other. Adenine is generally linked to thymine by two hydrogen bonds where as guanine is linked to cytosine by three hydrogen bonds.

(*v*) Since a purine is normally linked to pyrimidine, DNA has equal number of adenine and thymine as well as of guanine and cytosine residues.

(*vi*) There are 10 bases per turn of the helix.

Q. What is denaturation of DNA ?

Ans. Denaturation of DNA refers to the process of the separation of two strands when DNA solution is heated at a particular temperature. It is also called as melting of DNA.

Q. What is annealing ?

Ans. After denaturation, when the sample of DNA is slowly cooled, it, results in the formation of a double helix. This is due to complementary base-pairing.

This process is called as annealing or renaturation.

Q. What is hybridization ?

Ans. When one strand of the double helix DNA forms a complementary base pairing with RNA, such a double-stranded molecule is referred to as hybrid DNA and the process is referred to as hybridization.

Q. What is B-DNA ?

Ans. B-DNA is the most common form of DNA which has structural features as defined by Watson and Crick, *i.e.* it has two strands which are anti-parallel to each other and form a right-handed double helix. It has ten basepairs per turn and a pitch of 34 Å. Diameter of the double helix is 20 Å.

Q. What is A-DNA ?

Ans. A-DNA is a right-handed helix. It has 11 basepairs per turn with a pitch of 28 Å and an axial hole.

Q. What is Z-DNA ?

Ans. Z-DNA has complementary nucleotides with alternating purines and pyrimidines such as poly dGC or poly dAT. It has 12 basepairs per turn with a pitch of 45 Å units. There is a deep major groove but no identifiable minor groove and has a zig-zag conformation.

Q. What is mitochondrial DNA ?

Ans. The DNA present in the mitochondria in eukaryotes is referred to the mitochondrial DNA (mt DNA). It is a circular molecule and codes for various proteins which are required for oxidative phosphorylation within the mitochondria.

Q. What is RNA ?

Ans. RNA is a polynucleotide of ribonucleotides. There are three types of RNAs :

(*i*) mRNA,

(*ii*) tRNA (sRNA), and

(*iii*) rRNA.

Q. Discuss various differences between DNA and RNA ?

Ans. DNA and RNA differ form each other in several respects :

DNA	RNA
(*i*) DNA has deoxyribose.	RNA has ribose
(*ii*) DNA is a polymer of adenine, guanine, cytosine and thymine	RNA contains adenine, guanine, cytosine and uracil
(*iii*) DNA is a double-stranded right-handed helix.	RNA is generally a single-stranded molecule.
(*iv*) Number of adenine is equivalent to the number of thymine while that of guanine to cytosine.	Number of adenine need not be equal to uracil or of guanine to cytosine.
(*v*) DNA is of single type.	RNAs are of three types, each one of which has a specific function.

Q. What is mRNA ?

Ans. mRNA (messenger RNA) is the transcript of one of

the strands of DNA. It is a straight chain molecule whose bases are complementary to the bases present in the coding strand of DNA. It transfers the message from DNA to protein synthetic machinery. This message is present in the form of triplets.

Q. What is heterogeneous nuclear RNA (hn RNA) ?

Ans. hnRNA is a primary transcript produced in mammalian cells during transcription.

It is a large molecule which undergoes post-transcriptional modifications and is converted to an active mRNA.

In addition to sequences which code for protein (referred to as exons), hnRNA also has intervening noncoding sequences (referred to as introns). Introns are removed when hnRNA is processed and converted to mRNA.

Q. What is tRNA ?

Ans. tRNA (transfer RNA) is a type of RNA which transfers information from mRNA to protein synthetic machinery (ribosomes). Since it is present in the soluble fraction of the cell, it is also called as soluble RNA (sRNA).

It is an adaptor molecule and has a sequence of three nucleotides referred to as anticodon which is complementary to the sequence of three nucleotides (codon) on mRNA and determines the amino acid to be incorporated into the protein chain.

It has clover-leaf like structure due to extensive folding and intramolecular basepairing.

Q. Describe various features of tRNA.

Ans. Various features of tRNA are :

(*i*) Due to intramolecular folding, it has a double-stranded structure.

(*ii*) Each amino acid has atleast one specific tRNA.

(*iii*) All tRNAs, in addition to four nitrogenous bases, also contain several unusual bases, *e.g.* pseudouracil, hypoxanthine, etc.

(*iv*) Each tRNA has four arms with their specific functions. These arms are referred to as :

(*a*) **Amino acid Acceptor arm :** It has seven basepairs in a stem and terminates in CCA (in the $5'-3'$ direction). Adenosyl moiety acts as the acceptor of the amino acid.

(*b*) **Anticodon arm :** It has five basepairs and a loop. It also has a triplet (anticodon) which is complementary to the triplet (codon) on the mRNA.

Anticodon determines specificity of tRNA for carrying a particular amino acid.

(*c*) **DHU arm :** It has dihydrouracil, and

(*d*) **TψC arm :** It has thymine (T), pseudouracil and cytosine.

Q. What is rRNA ?

Ans. rRNA (ribosomal RNA) is a type of RNA which is found in ribosomes, in association with a large number of proteins. These RNA and protein particles are referred to as ribonucleoproteins.

Q. What are ribosomes ?

Ans. Ribosomes are granular particles found in the cytoplasm. They contain 80% of the total cellular RNA in association with several proteins. Both mRNA and tRNA interact on ribosomes and transfer information from DNA into a protein chain. Ribosomes are very complex molecules each one of which has two subunits, called as a large subunit and a small subunit.

Mammalian ribosomes have a sedimentation velocity of 80S. It has two subunits of 60S and 40S. The 60S molecule has three rRNAs (5S, 5.8S, 28S rRNA alongwith nearly fifty polypeptides). The 40S subunit of mammalian ribosome has 18S rRNA and about 30 proteins.

Q. What are the different sources of carbon and nitrogen in a purine ring ?

Ans.

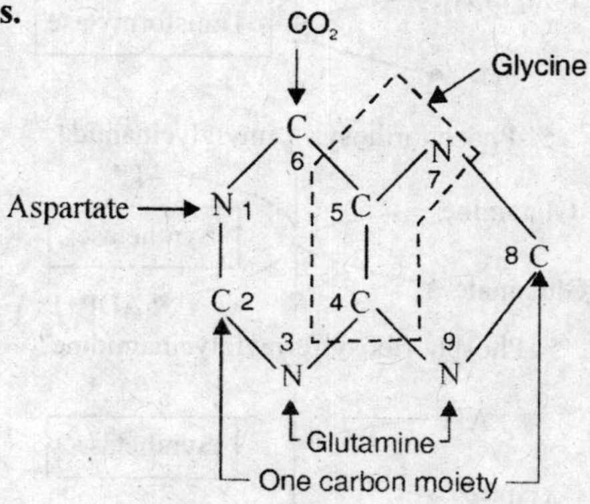

CO_2
Glycine
Aspartate
C 6
N 1
5 C
N 7
8 C
C 2
4 C
3 N
9 N
Glutamine
One carbon moiety

In a purine ring N–1 is derived from aspartate, C–2 and C–8 from one carbon moiety (N^{10} – formyl tetrahydrofolate) N–3 and N–9 from amide nitrogen of glutamine, C–6 from respiratory carbon-dioxide while C–4, C–5 and N–7 are contributed by glycine.

Q. Describe the process of De novo synthesis of purines ?

Ans. Purine nucleotides are synthesized from the various precursors by the process referred to as De novo synthesis. The starting material for a purine ring is phosphoribosylpyrophosphate (PRPP). It is synthesized by the enzyme PRPP synthetase which requires ribose-5-phosphate and ATP.

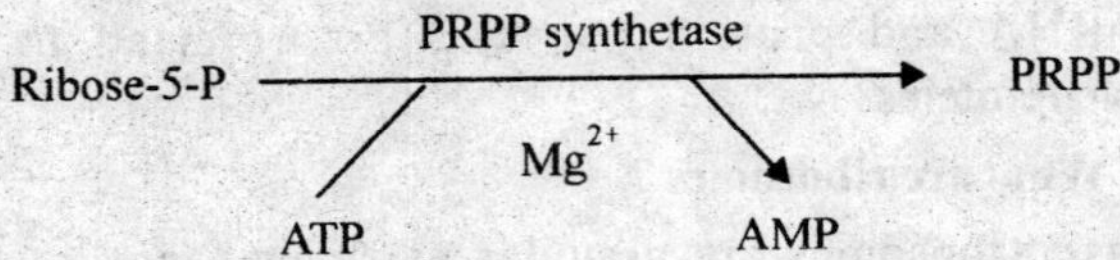

In this process first purine nucleotide which is synthesized is inosine monophosphate (IMP).

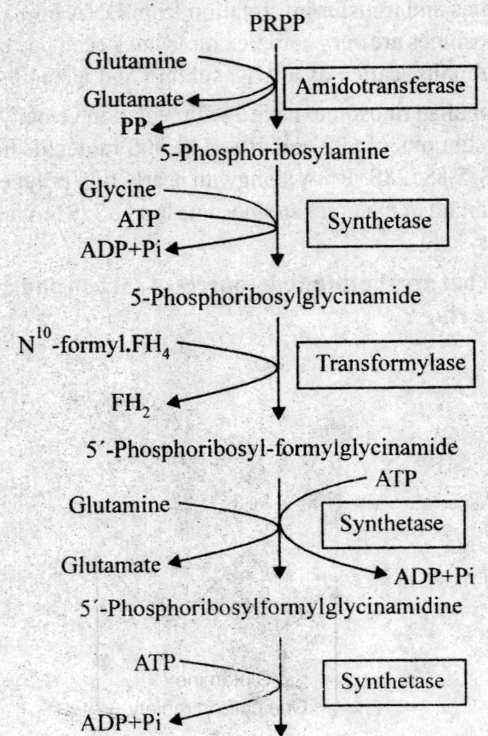

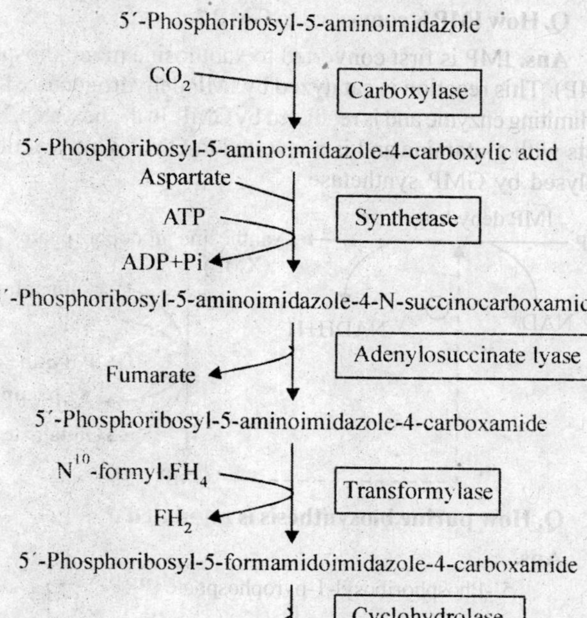

Q. How AMP is synthesized from IMP ?

Ans. IMP reacts with aspartate and forms adenylosuccinate (AMPS). This reaction is catalyzed by adenylosuccinate synthetase. In the next step, adenylosuccinase acts on AMPS and forms AMP.

AMP acts as competitive inhibitor for adenyloscuccinate synthetase which is a rate limiting enzyme.

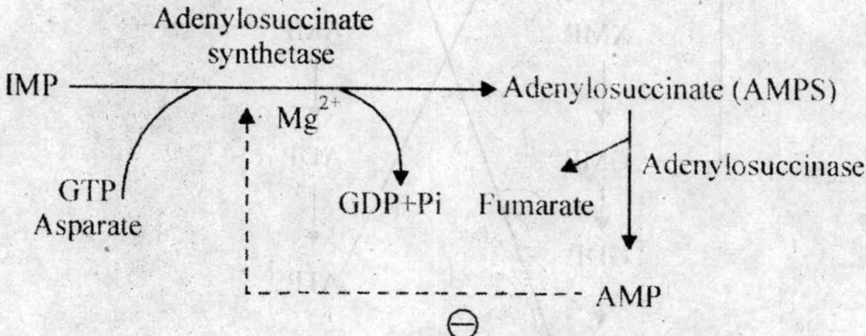

Q. How IMP is converted to GMP ?

Ans. IMP is first converted to xanthosine monophosphate (XMP). This reaction is catalyzed by IMP dehydrogenase. It is a rate limiting enzyme and is regulated by GMP. In the next step, XMP reacts with glutamine and is converted to GMP. This reaction is catalysed by GMP synthetase.

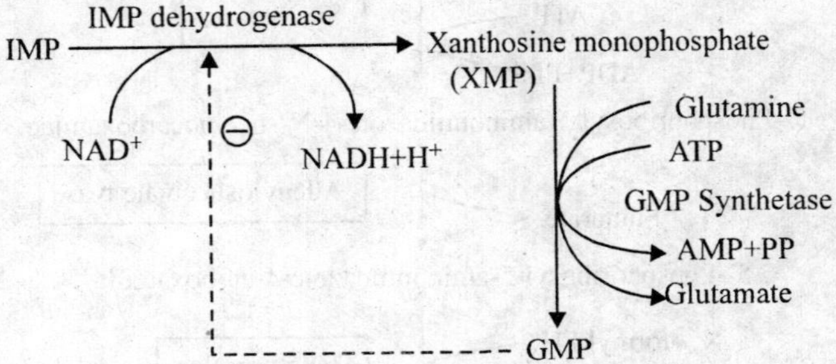

Q. How purine biosynthesis is regulated ?

Ans.

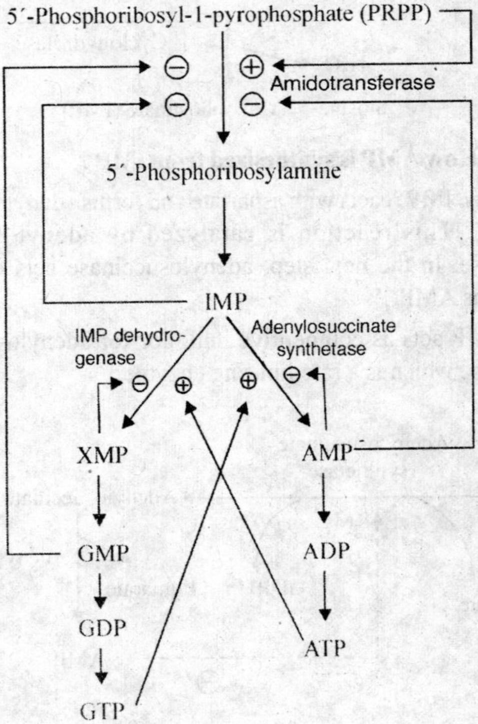

5′-Phosphoribosyl-1-pyrophosphate (PRPP)

Amidotransferase

5′-Phosphoribosylamine

IMP

IMP dehydrogenase

Adenylosuccinate synthetase

XMP

AMP

GMP

ADP

GDP

ATP

GTP

Biosynthesis of IMP is regulated by amidotransferase. It is a committed step and is allosterically inhibited by the end products, *i.e.* IMP, GMP and AMP. This enzyme is stimulated by PRPP.

Biosynthesis of AMP from IMP is also inhibited by the end product AMP. Adenylosuccinate synthetase is competitively inhibited by AMP and stimulated by GTP.

IMP dehydrogenase is inhibited by GMP whereas ATP stimulates this reaction.

Q. Describe the salvage pathway of purine biosynthesis ?

Ans. Salvage pathway refers to the synthesis of nucleotides from free purine bases and nucleosides.

Free nitrogenous bases can be converted to their nucleotides with the utilization of phosphoribosyl pyrophosphate (PRPP).

Adenine is converted to AMP, by adenosine phosphoribosyltransferase.

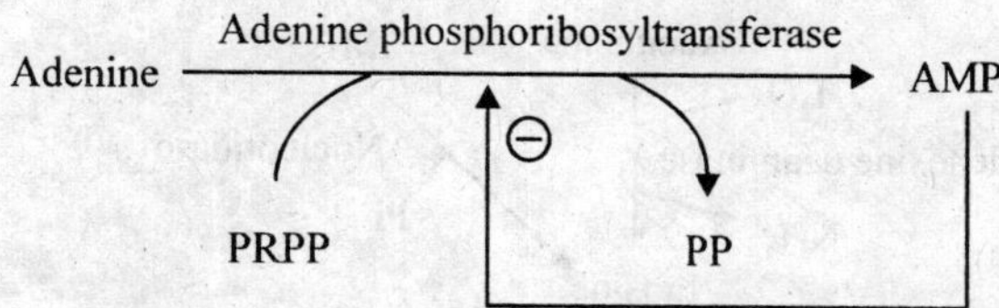

On the other hand, hypoxanthine-guanine phosphoribosyltransferase (HGPRTase) converts both hypoxanthine as well as guanine to their respective nucleotides *i.e.* IMP and GMP, respectively.

Both the enzymes are inhibited by their end products.

Nucleosides can also be salvaged to nucleotides by nucleoside kinases.

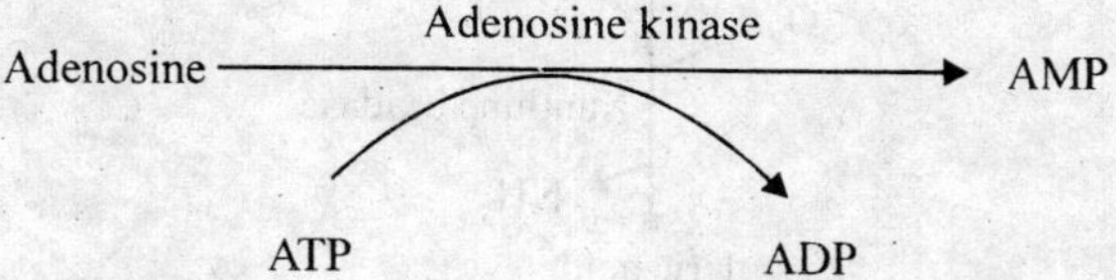

These reactions conserve energy and are important in some of the cells such as erythrocytes which lack enzymes for De novo synthesis.

Q. What is Lesch-Nyhan syndrome ?

Ans. Lesch-Nyhan syndrome refers to the deficiency of the enzyme HGPRTase. As a result of it, hypoxanthine and guanine can not be utilized by the salvage pathway. This, in turn, results in increased concentration of PRPP and promotes De novo synthesis of purines. There is also excessive catabolism of purines (hypoxanthine and guanine), leading to hyperuricemia.

Q. How are purines catabolized ?

Ans. Purines are catabolized to uric acid.

AMP is converted to adenosine by a nucleotidase. Thereafter, adenosine deaminase removes ammonia and forms inosine.

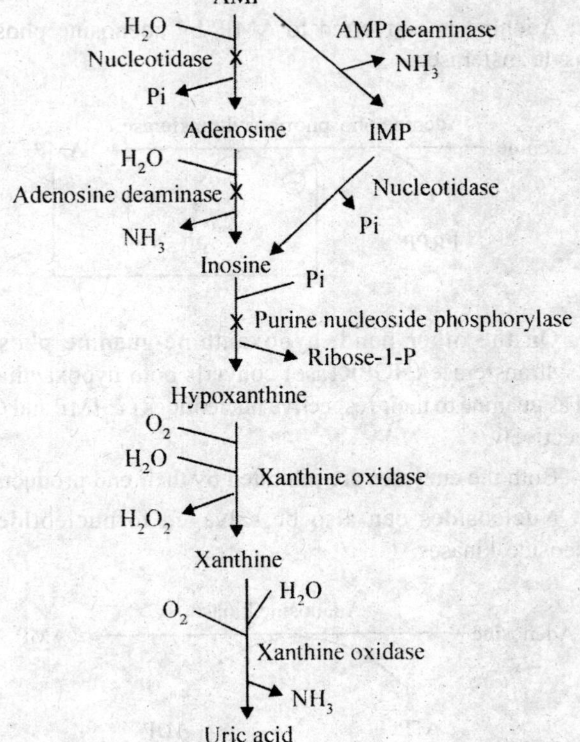

Alternatively, AMP deaminase can also remove ammonia from AMP and converts it to IMP which is subsequentely converted to inosine by a nucleotidase. Thereafter, inosine is converted to hypoxanthine by purine nucleoside phosphorylase. Hypoxanthine is converted to xanthine by xanthine oxidase which is further oxidized to uric acid.

GMP is converted to guanosine by a nucleotidase. Guanosine is then converted to guanine by purine nucleoside phosphorylase and finally to xanthine by guaninase.

Xanthine is subsequentely oxidized to uric acid by xanthine oxidase.

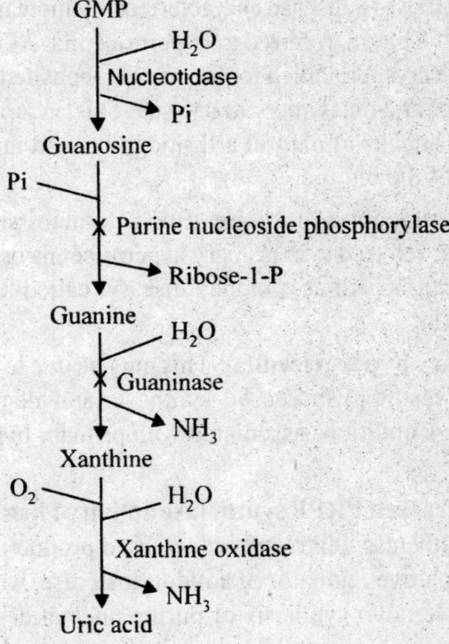

Q. Describe immuno-deficiency diseases which are associated with the degradation of purines ?

Ans. There are two disorders which are associated with the degradation of purine nucleosides –

1. Purine nucleoside phosphorylase deficiency (PNPD) : Purine nucleoside phosphorylase converts a purine nucleoside to free nitrogenous base. Its deficiency is associated with defective T-cell functions. However, B-cell functions are normal.

2. Severe combined immuno-deficiency (SCID) : It is due to the deficiency of adenosine deaminase which converts adenosine to inosine. It results in combined deficiency of, both T-cell as well as B-cell functions. Such children are more prone to infections which may result in their death within two years of age. Recently, gene therapy for ADA gene has been used in the treatment of SCID.

In both the conditions, there are increased levels of purine nucleosides and nucleotides alongwith decrease in uric acid.

Q. What is gout ?

Ans. Gout is a disease characterized by increased levels of uric acid in the blood, referred to hyperuricemia. As a result of it, sodium urate crystals, called tophi, may be deposited in the joints of extremities and renal interstitial tissue. This, in turn, may result in recurrent attacks of painful arthritic joints and inflammation, most often of the big toe.

Hyperuricemia may be due to various metabolic abnormalities referred to as primary hyperuricemia or due to some secondary cause such as a kidney disease, called as secondary hyperuricemia.

Primary hyperuricemia : This may be due to a metabolic defect that result in increased synthesis and degradation of nucleosides. Conditions associated with primary hyperuricemia include –

(*i*) **Increased PRPP synthetase activity :** Increased levels of PRPP synthetase will result in increased production of PRPP which is a positive regulator of amidotransferase, which is a key enzyme in De novo synthesis of purine nucleotides. Increased synthesis of purines, in turn, results in their increased catabolism, leading to hyperuricemia.

(*ii*) **Lesch-Nyhan syndrome :** Deficiency of the enzyme HGPRTase reduces purine salvage pathway and results in the accumulation of hypoxanthine and guanine, and their subsequent degradation to uric acid.

(*iii*) **Von Gierke's disease (Glycogen storage disease type I) :** It is due to the deficiency of glucose-6-phosphatase. Increased level of glucose-6-phosphate results in increased production of ribose-5-phosphatase. Increased availability of ribose-5-phosphate, in turn, stimulates the production of PRPP.

In addition, lactic acidosis also elevates renal threshold for urate.

Secondary hyperuricemia : It occurs as a result of some disease other than a metabolic abnormality such as a kidney disease or a disease associated with increased cell turnover, *e.g.* leukemia or psoriasis.

Q. What are the sources of carbons and nitrogens in a pyrimidine ring ?

Ans. In a pyrimidine ring N_1, C_4, C_5 and C_6 are derived from aspartate, C_2 from respiratory CO_2 and N_3 from amide nitrogen of glutamine.

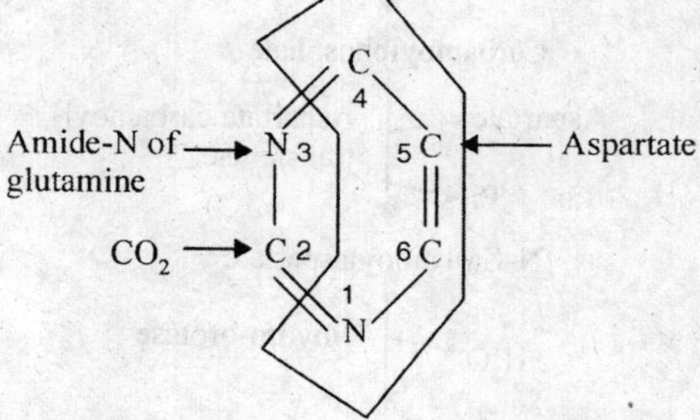

Q. Describe the process of De novo synthesis of pyrimidines.

Ans. A pyrimidine ring is first synthesized which is thereafter converted to UMP. UMP serves as a precursor of CMP as well as TMP.

Biosynthesis of UMP : (*i*) In the first step, CO_2 (as HCO_3^-) reacts with amide-nitrogen of glutamine and forms carbamoyl-phosphate.

This reaction is catalysed by carbamoylphosphate synthetase II, in the cytosol.

(*ii*) Carbamoylphosphate reacts with aspartate and form N-carbamoylaspartate. This is a committed step in pyrimidine biosynthesis and is catalyzed by aspartate transcarbamoylase.

(*iii*) Carbamoylaspartate is converted to dihydro-orotate,

by dihydroorotase.

(*iv*) Dihydro-orotate is converted to orotate by dihyro-orotate dehydrogenase.

(*v*) Orotate in converted to orotidine monophosphate (OMP) with the transfer of ribose-5-phosphate, from PRPP. This reaction is catalyzed by orotate phosphoribosyltransferase.

(*vi*) Finally, OMP is decarboxylated to UMP, by OMP decarboxylase.

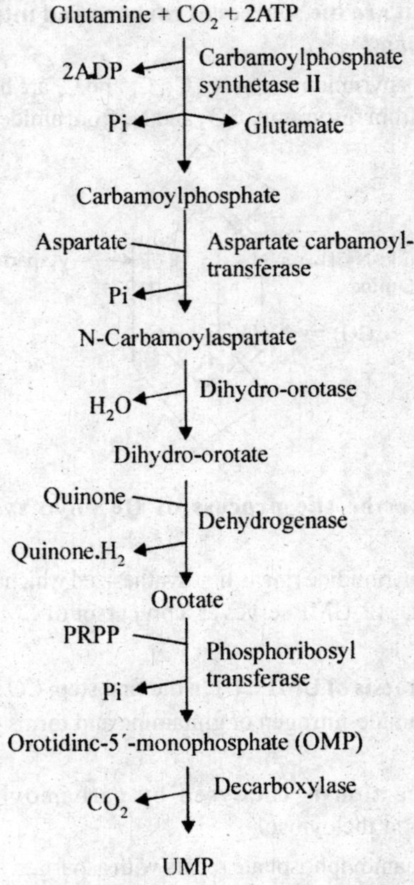

Q. What is oroticaciduria ?

Ans. Oroticaciduria is a genetic disorder, due to the lack of orotate phosphoribosyltransferase and/or OMP-decarboxylase. Deficiency of any of these two enzymes will result in the accumulation of orotate and its increased excretion in the urine. This is characterized by growth retardation, due to reduced synthesis of pyrimidines.

Q. How a nucleoside monophosphate can be converted to di- and -triphosphates ?

Ans. Nucleoside monophosphate (NMP) can be converted to di- and -triphosphates, by sequential phosphorylation, by ATP. This reaction is catalyzed by kinases.

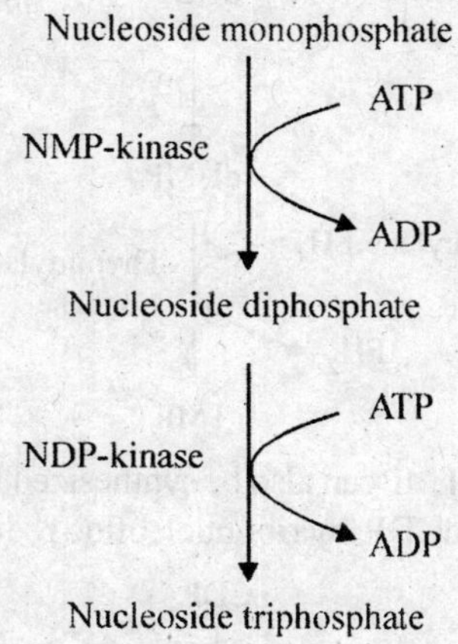

Q. How UMP is converted to CMP ?

Ans. UMP is first converted to UDP and then to UTP. Thereafter, UTP is converted to CTP by CTP synthetase. Glutamine is used as a donor of the amino group. CTP in turn, is first converted to CDP and then to CMP, by sequential dephosphorylation.

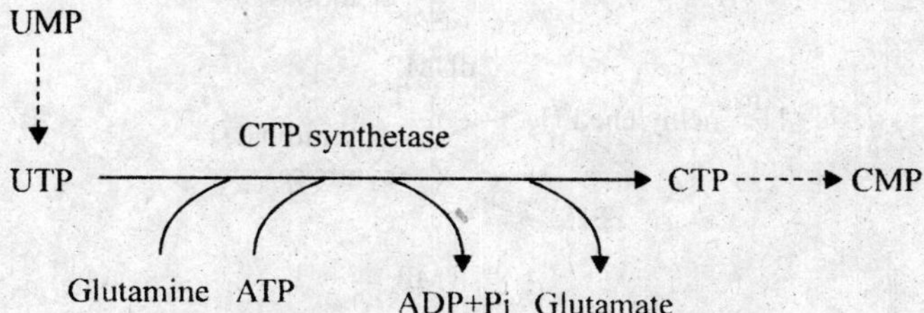

Q. How is TMP synthesized ?

Ans. TMP can be synthesized from either CDP or UDP.

UDP is first oxidized to dUDP by ribonucleotide reductase. In the next step, deoxy -UDP is converted to dUMP. dUMP is then converted to TMP by thymidylate synthase, by using N^5, N^{10}-methylenetetrahydrofolate as a source of the methyl group.

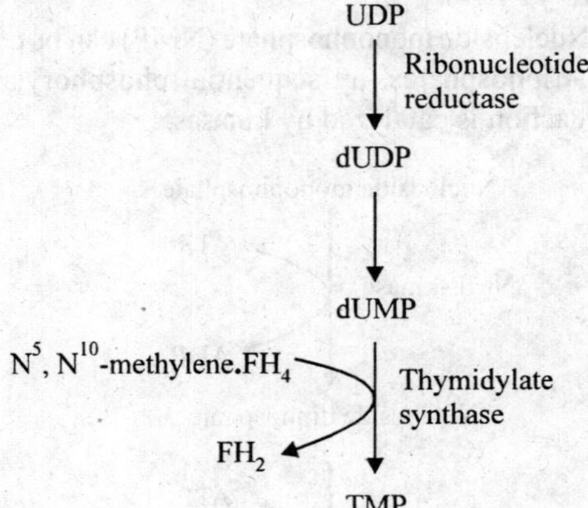

Alternatively, TMP can also be synthesized from CDP. CDP is first converted to dCDP, by ribonucleotide reductase. dCDP is

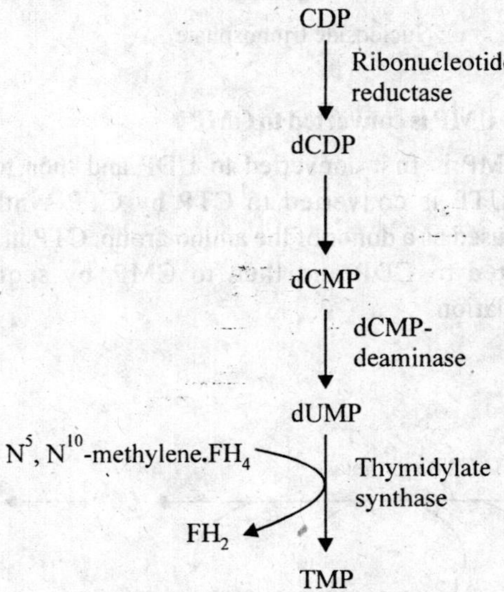

then converted to dCMP. dCMP is converted to dUMP by deaminase. Finally, dUMP is methylated to form dTMP. This reaction is catalyzed by thymidylate synthase, using N^5, N^{10}-methylenetetrahydrofolate as a donor of the methyl group.

Q. What are suicide inhibitors ?

Ans. Substances such as 5-fluorodeoxy-UMP (FdUMP) are used as a substrate and inactivates the enzymes after undergoing some of the catalytic reaction, *i.e.* FdUMP binds to the enzyme thymidylate synthase and forms enzyme-FdUMP-FH_4 ternary complex which can not form the product. Since, these inhibitors cause the enzyme to commit suicide, such substances are called as suicide inhibitors. These are used as anticancer drugs.

Q. What are antifolates ?

Ans. Antimetabolates, such as amethopterine (methotrexate) and aminopterine are the structural analogues of dihydrofolate and inhibit dihydrofolate reductase. This, in turn, blocks the regeneration of tetrahydrofolate and inhibits the synthesis of dTMP. These substances are thus called as antifolates.

Q. How pyrimidine biosynthesis is regulated ?

Ans. Biosynthesis of pyrimidine nucleotides is inhibited in eukaryotes at three steps :

(*i*) Carbamoylphosphate synthase - II : This enzyme is inhibited by CTP, and activated by ATP and PRPP.

(*ii*) OMP decarboxylase : It is competitively inhibited by UMP.

(*ii*) CTP synthase : It is inhibited by CTP, by feed back inhibition.

Q. Describe the salvage pathway for pyrimidines ?

Ans. Orotate, uracil and thymine (but not cytosine) can be salvaged and converted to the corresponding nucleotides, by pyrimidine phosphoribosyltransferase. The enzyme uses PRPP as a source of ribose-5-phosphate.

$$\text{Uracil} \xrightarrow[\text{Phosphoribosyltransferase}]{} \text{UMP}$$

Q. How are pyrimidine nucleotides catabolized ?

Ans. 1. Pyrimidine nucleotides are converted to the

corresponding nucleosides by phosphatases.

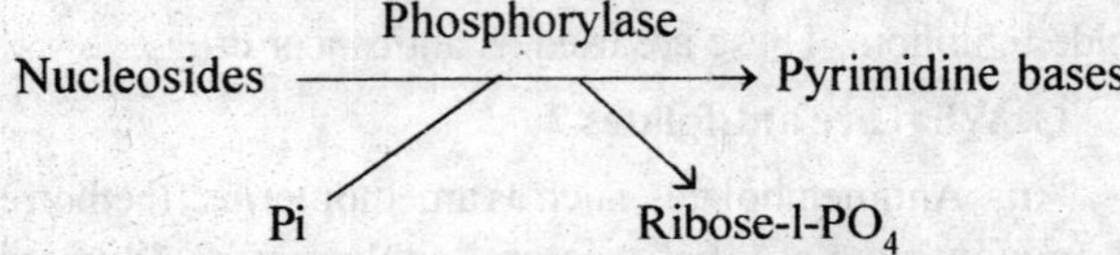

2. Cytidine and deoxycytidine are deaminated, by nucleoside deaminase, to uridine and deoxyuridine, respectively.

Deaminase

Cytidine/d-Cytidine ⟶ Uridine/d-Uridine

3. Nucleosides (uridine/d-uridine and thymidine) are converted to free nitrogenous bases by phosphorylases.

Phosphorylase

Nucleosides ⟶ Pyrimidine bases

Pi Ribose-l-PO_4

4. Nitrogenous bases (uracil/thymine) are thereafter converted to the excretory waste products, *i.e.* β-alanine and β-amino-isobutyrate, respectively.

Q. How deoxyribonucleotides are synthesized ?

Ans. In the first step, deoxyribonucleotides are synthesized by the reduction of the ribose unit of the ribonucleotide-diphosphates, by ribonucleotide reductase.

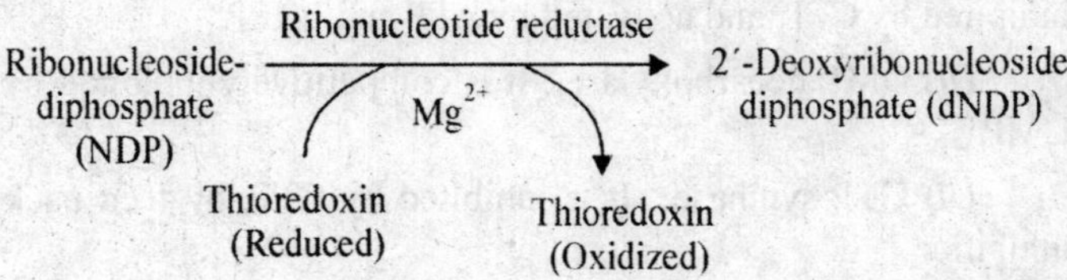

2. Subsequently, deoxynucleosidediphosphates are converted to deoxynucleosidetriphosphates by kinase which requires ATP

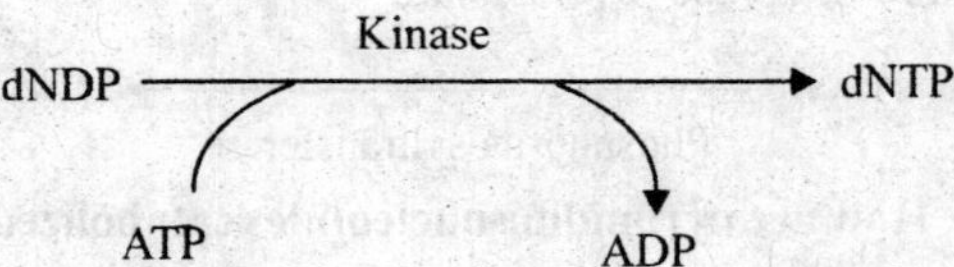

Q. What is replication ?

Ans. The process of DNA synthesis is referred to as replication. It is carried out by DNA dependent DNA polymerases, also referred to as DNA polymerases.

Q. Explain the process of semiconservative replication.

Ans. The process of DNA replication where each strand of the double-stranded DNA serves as a template for the synthesis of a new strand is called semiconservative replication. During each round of replication the double stranded molecule has a parental strand and a newly synthesized daughter strand.

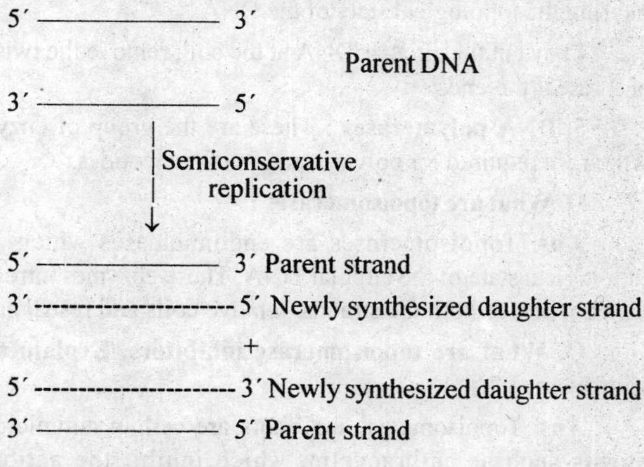

Q. Define origin of replication ?

Ans. Origin of replication is the site on the chromosome where DNA replication starts. This site is rich in A-T sequence.

In E.coli this site is referred to as OriC.

In eukaryotes, replication starts at A-T rich region at several locations, simultaneously, which are referred to as autonomously replicating sequence elements (ARS elements).

Q. Name different proteins and enzymes required for the process of replication ?

Ans. For the process of DNA replication following proteins and enzymes are required :

1. **DnaA protein :** It is also called as origin-binding protein. Several molecules of DnaA bind at the origin of replication which is rich in A and T basepairs.

2. **Single-strand DNA binding protein :** After unwinding of the double-stranded DNA, single-strand DNA binding protein (SSB protein) binds to both the strands separately so as to prevent their rejoining.

3. **Helicase :** DNA helicase binds at the junction of the two single-strands and the double stranded molecule, after unwinding of the two strand. DNA helicase further helps in the unwinding of the DNA.

4. **Topoisomerase :** Topoisomerases are endonucleases which remove coiling from the circular DNA. Thus they help in altering the topological state of the DNA.

They cut the circular DNA at the coil, remove the twisting and reseal the ends.

5. **DNA polymerases :** These are the group of enzymes which are required for polymerization of nucleotides.

Q. What are topoisomerases ?

Ans. Topoisomerases are endonucleases which alter topological state of the circular DNA. These enzymes introduce nicks in one or both the strands, remove coils and reseal ends.

Q. What are topoisomerase inhibitors. Explain their significance ?

Ans. Topoisomerase inhibitors are certain antimicrobial agents such as anthracyclin, which inhibit the action of topoisomerase.

These drugs inhibit resealing of the nicked strands of DNA, and thus partially inhibit the action of topoisomerases. Topoisomerase inhibitors are used in the treatment of certain hematological malignancies, such as leukemia.

Q. What are DNA polymerases. Differentiate between prokaryotic and eukaryotic enzymes ?

Ans. DNA polymerases are enzymes which catalyze the synthesis of a complementary strand, on the single-stranded DNA template. Three types of DNA polymerases are found in prokaryotes :

1. **DNA Polymerase I :** It has three activities, *i.e.* $5' \rightarrow 3'$ polymerase activity, and $3' \rightarrow 5'$ as well as $5' \rightarrow 3'$ exonuclease activity.

2. **DNA Polymerase II :** It has 5′ → 3′ polymerase and 3′ → 5′ exonuclease activity but, lacks 5′ → 3′ exonuclease activity. It mainly participates in DNA repair.

3. **DNA Polymerase III :** It is required in De novo synthesis of DNA. It can polymerise DNA as well as edit mistakes.

It is a more complex enzyme and consists of large number of subunits. It has a catalytic core which consists of three subunits (α, ε and θ) besides seven other subunits, out of which five form a complex with the catalytic core whereas other two (β and γ subunits) form a clamp around the template.

There are five types of polymerases in eukaryotes :

1. **DNA Polymerase α :** It has 5′ → 3′ polymerase activity. It is also called as the lagging strand replicase and is concerned with initiation of replication.

2. **DNA Polymerase δ :** It has both 5′ → 3′ as well as 3′ → 5′ polymerase activity. It however, requires another protein referred to as **proliferating cell nuclear antigen** (PCNA), for high processivity.

The holoenzyme (polymerase δ-PCNA complex) is also referred to the leading strand replicase.

3. **DNA Polymerase ε :** It has 5′ → 3′ polymerase activity as well as 3′ → 5′ exonuclease activity. It is highly processive even in the absence of **PCNA.**

It is mainly involved in repairing the damaged nuclear DNA.

4. **DNA Polymerase β :** It is a small polypeptide and participates in the repair of the damaged nuclear DNA. However, its biological function is still unknown.

5. **DNA Polymerase γ :** It is exclusively found in the mitochondria and involves in mitochondrial DNA replication.

Q. What is reverse transcriptase ?

Ans. Reverse transcriptase is an RNA dependent DNA polymerase. It catalyzes synthesis of the double-stranded DNA from the single-stranded RNA template in retroviruses such as HIV.

Q. Describe the features of DNA Polymerase I.

Ans. DNA polymerase I is a prokaryotic enzyme which has an important role in DNA replication as well as its repair in E.coli. It has 5′ → 3′ polymerase activity alongwith 3′ → 5′ as well as 5′

→ 3′ exonuclease activity. Due to its proof reading action, it is important in the repair of the damaged DNA.

DNA Polymerase I is a single polypeptide constituting of 2 fragments, *i.e.* a large fragment and a small fragment. The large fragment is referred to as Klenow fragment. It has both 5′ → 3′ polymerase and 3′ → 5′ exonuclease activity.

The small fragment has 5′ → 3′ exonuclease activity.

Q. Describe features of DNA polymerase III.

Ans. DNA polymerase III is a replicase which is actively involved in the De novo synthesis of DNA as well as to edit mistakes. The enzyme however, lacks nick translation activity. It has 5′ → 3′ polymerase activity and 3′ → 5′ exonuclease activity.

Structurally, DNA polymerase III is a complex enzyme and consists of several subunits. It has catalytic core which consists of three subunits, referred to as α, ε and θ. Its α subunit has DNA polymerase activity while ε subunit has 3′ → 5′ exonuclease activity. Besides, there are seven other subunits which form a complex with the catalytic core of enzyme while two other subunits form a clamp around the template. These are referred to as β and γ subunits. Its β subunit acts as a clamp and promotes processivity, whereas γ subunit acts as a clamp holder.

Q. Describe features of DNA polymerase δ.

Ans. DNA polymerase δ is a eukaryotic enzyme that catalyses DNA replication with high processivity after its association with proliferating cell nuclear antigen (PCNA). Polymerase δ-PCNA complex is also referred to as a leading strand replicase since it helps in replication of the leading strand.

Q. What is replication bubble ?

Ans. Replication bubble, also refers to as replication eye is the formation of bubble like structure during the initiation of DNA replication.

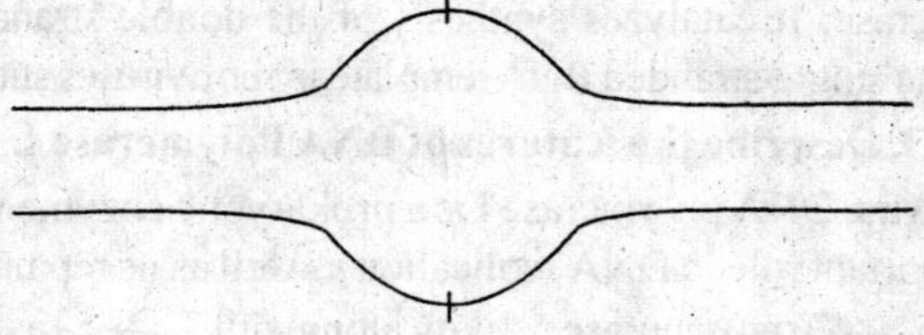

Q. What is replication fork ?

Ans. Replication fork is a branch point in a replication bubble at which DNA synthesis starts. Formation of a replication fork initiates stepwise assembly of all the enzymes and other proteins that are necessary to initiate the process of DNA replication.

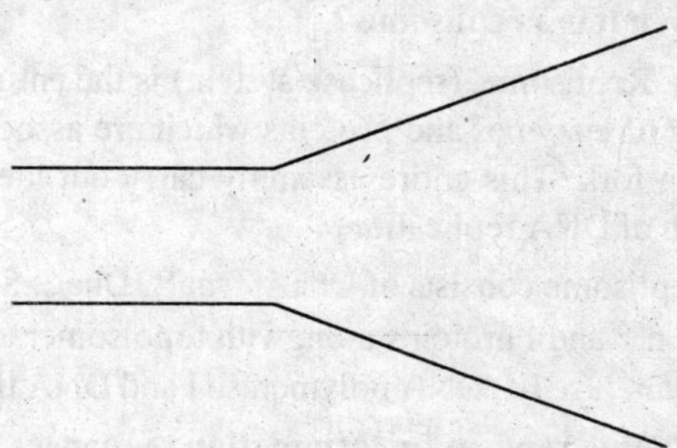

Q. What are Okazaki fragments ?

Ans. Okazaki fragments are small DNA fragments, also referred to as nascent DNA segments or precursor fragments, which are synthesized discontinuously on the lagging strand, during the process of DNA replication.

They may vary in size from nearly 100 to 200 deoxyribonucleotides in eukaryotes to nearly 1000 to 2000 deoxyribonucleotides in prokaryotes.

Q. What is a primosome ?

Ans. A primosome is a complex of primase enzyme and the prepriming assembly which consists of several proteins that are necessary for the synthesis of RNA primer.

Prepriming assembly comprises of DnaB, DnaC, n, n′, n′′ and i proteins. Primosome searches DNA for the specific sequences and allows the primase to synthesize RNA primer in the 5′ → 3′ direction.

Q. What is an RNA primer ?

Ans. RNA primer is a short stretch of ribonucleotides that is synthesized by the primase-prepriming assembly on a DNA template. These are 10 to 60 ribonucleotides long and provide free 3′– OH group for the initiation of DNA synthesis. These are removed and replaced by the deoxyribonucleotides before daughter DNA strand is released.

3′ ———————— 5′ DNA template

5′ - - - - - - - - - - ————→ 3′

Q. What is a replisome ?

Ans. Replisome, (replicase system) is the entire assembly consisting of enzymes and proteins which are associated at the replication fork. This entire assembly carry out the process of elongation of DNA replication.

A replisome consists of DnaA, DnaB, DnaC, SSB protein, and n, n′, n′′ and i proteins alongwith topoisomerase, primase, DNA polymerase III, DNA polymerase I and DNA ligase.

Q. Define replication termination sequences ?

Ans. Replication termination sequences (TER sites) are the sequences that result in termination of DNA replication.

These are G.T rich sequences and located directly opposite to the site of origin of replication.

In E.coli, there are two TER sites which are referred to as TER-E, D and A, and TER-G, F, B and C. These are located nearly 100 kilobase on either side of the meeting point.

Two replication forks, that initiates bidirectional DNA replication at the origin, meet at the oppositely facing TER sites.

Q. Describe the process of DNA replication in prokaryotes ?

Ans. Replication of DNA takes place in three steps :

1. Initiation,

2. Elongation and

3. Termination.

1. Initiation : DNA replication begins at the point referred to as origin of replication which is rich in A-T sequences. DnaA protein binds at the origin alongwith another protein referred to as DnaC. This binding of DnaA results in local melting of DNA in the A-T rich region. There is also binding of DnaB protein which, in turn, creates the formation of replication fork. Topoisomerase removes the supercoils which are formed during the unwinding of DNA. With the advancement of helicase two single-stranded regions are formed. These are covered with the single-stranded DNA binding protein.

Replication starts on the two antiparallel single-stranded DNA templates, simultaneously, at two different replication forks but in opposite directions. On one of the strands, there is continuous synthesis of the new strand in the direction of the movement of the fork. This strand is referred to as the leading strand.

On the other strand, DNA synthesis occurs, discontinuously, in smaller fragment. These fragments are referred to as Okazaki fragments while this strand is designated as the lagging strand.

Replication takes place in the 5′ → 3′ direction on both the strands. The process of replication begins with the synthesis of RNA primer. At this point, besides DnaB and DnaC, binding of n, n′, n′′ and i proteins, and primase (with an RNA polymerase activity) also occurs. This whole complex is referred to as primosome.

Primosome recognizes TGGT and GGGT sequences and interacts with a template at the replication fork to begin the synthesis of RNA primer. SSB proteins are displaced and primase synthesizes the primer.

Only one RNA primer is synthesized on the leading strand while several RNA primers are synthesized on the lagging strand.

2. Elongation : DNA polymerase III, DNA polymerase I and DNA ligase gets associated with the primosome, and form replicase unit referred to as replisome. Replicase extends the RNA primer, with the incorporation of deoxyribonucleotides and leads to the synthesis of the new DNA strand.

DNA replication takes place in 5′ – 3′ direction, on both the templates. The hollow enzyme moves as a single unit in the 5′ to 3′ direction along the leading strand. On the lagging strand, after completing the synthesis of one Okazaki fragment, it gets relocated to a new primer and extends the synthesis of the other Okazaki fragment. This, in turn, results in the continuous synthesis of a leading strand and several fragments, which are separated from each other, on the lagging strand.

The DNA polymerase I catalyzes replacement of the ribonucleotides, of the RNA primer, with the deoxyribonucleotides while the nicks, present within the lagging strand, are sealed by DNA ligase.

3. Termination : Termination of DNA replication takes place after the binding of a contra-helicase at the TER site, located directly opposite to the origin of replication and has G-T rich sequences. Contra-helicase is a protein referred to as Tus protein (a terminator utilization substance). Tus protein binds at the G-T rich region on the TER site and prevents the advancement of the replication fork, by arresting the action of helicase (DnaB protein). Finally, unlinking of parent strands, from the daughter strands, is catalyzed by topoisomerase.

Q. How process of DNA replication in prokaryotes differs from that of eukaryotes.

Ans. DNA replication in eukaryotes is more complex than prokaryotes. Differences in DNA replication in prokaryotes and eukaryotes –

Prokaryotes	Eukaryotes
1. Replication begins at a single site	Replication begins at multiple sites, simultaneously
2. Origin of replication is rich in A-T nucleotides. It is referred to as oriC in a bacterial chromosome.	Origin of replication is referred to as autonomous replicating sequence (ARS) which has A-T rich central region consisting of 11 bp element (ARS-consensus sequence).
3. The two strands, after their separation, are covered with the single-strand DNA-binding protein.	DNA synthesis requires replication protein-A (replication factor-A)
4. Three types of DNA polymerases are found, *i.e.* DNA polymerase I, II and III.	There are five different types of DNA polymerases . These are referred to as polymerase α, β, γ, δ and ε
5. Helicase activity is present in DnaB	Helicase activity is associated with DNA polymerase δ
6. Synthesis of RNA primer is associated with primase which has RNA polymerase activity	Primase activity is associated with DNA polymerase α

7. Synthesis of, both, the leading strand and the lagging strand is carried out by two similar but distinct subunits of DNA polymerase III	Leading strand is synthesized by DNA polymerase δ while synthesis of the lagging strand is carried out by DNA polymerase δ
8. Okazaki fragments are 1000 to 2000 nucleotides long	Okazaki fragments are 100 to 200 nucleotides long
9. DNA replication occurs throughout the cell cycle	DNA synthesis is confined to S phase of the cell cycle only.
10. Initiation of DNA synthesis takes place after the binding of DnaA protein at the origin of replication	Initiation of DNA synthesis is regulated by cyclins and cyclin dependent kinases
11. Separation of daughter strands from parent strands is caused by telomerase	DNA is packed alongwith histones which are also synthesized, simultaneously
12. No telomeres are present	Ends of chromosomes have unusual structure called telomeres.

Q. What are telomeres ?

Ans. Telomeres are 1000 or more nucleotides long tandem repeats of G-T rich sequences that are present at the 3′ end of each chromosome. A telomere also has a 12 to 16 nucleotides long single-strand which serves as a template for the primer, for the synthesis of the final Okazaki fragment on the lagging strand.

In most of the eukaryotes, replication of the telomere is carried out by a specific reverse transcriptase referred to as telomerase.

Telomere repeats gradually decrease in number with aging.

Q. How a defect in DNA synthesis is repaired ?

Ans. A segment of the damaged DNA is repaired by the process referred to as excision repair. This may include base excision repair or nucleotide excision repair.

Base excision repair : Bases are hydrolytically removed by glycosylases. This, in turn, results in the production of a site called apurinic-apyrimidinic site (A-P site). An AP endonuclease nicks the phosphodiester back-bone at A-P site and results in excision of the sugar-phosphate residue. Gap is filled by DNA polymerase I while the ends are sealed by DNA ligase.

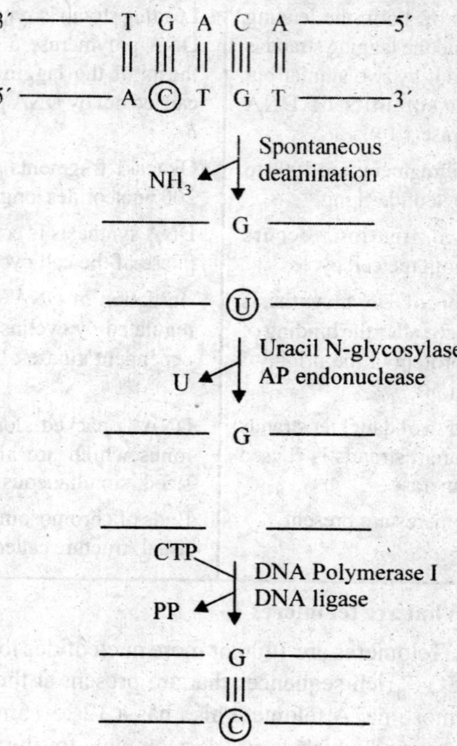

Nucleotide excision repair : The damaged strand is excised by an endonuclease system, referred to as exci nuclease'. The defective segment is removed and repaired by DNA polymerase I and DNA ligase.

Q. What is xeroderma pigmentosum ?

Ans. Xeroderma pigmentosum is characterized by neurological abnormalities and skin cancer. Patients exhibit sunlight induced photodermatitis which, in turn, may lead to skin ulceration and cancer.

It is due to a defect in the DNA repair system. UV light-induces damage to the DNA segment such as thymine-thymine dimerization, which is normally repaired by nucleotide excision repair. Excinuclease in eukaryotes comprises of several subunits. The 3′ and 5′ nick is introduced and an oligonucleotide is excised. The gap is filled by polymerases γ and ε, and finally the nick is sealed by DNA ligase.

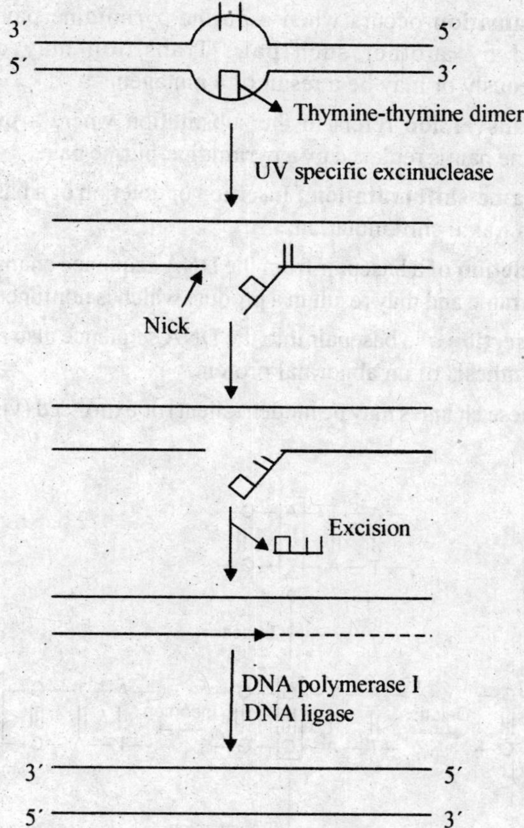

Q. What is mutation ? Describe various types of mutations.

Ans. Mutation can be defined as a permanent change in the DNA segment of a gene which, in turn, may be expressed as a phenotypic change in the organism. Mutation may occur due to error in replication, UV or ionising radiation or by a chemical mutagen.

Mutations may be classified as :

1. Point mutation, and

2. Frame-shift mutation.

Point mutation : Point mutation may occur due to a change in the single basepair in the DNA. It may be a result of transition or transversion.

Transition occurs when a purine-pyrimidine pair gets replaced by another such pair. Transition may occur spontaneously or may be a result of a mutagen.

Transversion refers to the substitution where a purine-pyrimidine pair is replaced by a pyrimidine-purine pair.

Frame-shift mutation : Insertion or deletion of a basepair results in frame-shift mutation.

Deletion of a basepair from the DNA sequence changes its reading frame and may result in a product which is nonfunctional.

Insertion of a basepair into the DNA sequence also results in the synthesis of an abnormal protein.

These changes may be hidden (silent) or expressed (visible).

Q. What is a silent mutation ?

Ans. Any change in a base, in the third position of the codon, that may not affect the specified amino acid, *e.g.* UUA to UUG, as both codes for leucine. Such mutations are called silent mutations.

Q. What are visible mutations ?

Ans. A mutation which may result in the expression of an abnormal gene product, *e.g.* a change from GAA or GAG to GUA or GUG, results in the incorporation of valine instead of glutamate causing sickle cell anemia. Such mutations are called visible mutations.

Q. What is transcription ?

Ans. Transcription is the process by which sequences of nucleotides in DNA are transcribed into the sequence of ribonucleotides into RNA.

It is a strand selective process since during transcription only one of the strands of the DNA, that is referred to as the template strand, is transcribed. It is catalyzed by the enzyme RNA polymerase.

Q. What is reverse transcription ?

Ans. Reverse transcription is the process of transfer of genetic information in RNA containing viruses. Genetic material in these viruses is RNA which is first transcribed into DNA and thereafter this DNA is further transcribed for the synthesis of RNA.

In these retroviruses this process of reverse transcription is catalyzed by the enzyme referred to as reverse transcriptase.

Q. Write note on upstream and downstream sequences ?

Ans.

Upstream refer to the sequence of nucleotides prior to the transcription start site. These nucleotides are assigned negative numbers and are present towards the 5′ end.

Downstream refer to the sequence of nucleotides after the transcription start site. These nucleotides are assigned positive numbers and are present towards the 3′ end. The first nucleotide at the transcription start site is written as + 1.

Q. What is a promotor ?

Ans. Promotors are the sequences of nucleotides at the specific sites on the DNA template where RNA polymerase binds. Promotor region, in prokaryotes consists of 45 bp upstream to 10 bp downstream. It has two short nucleotide sequences which are highly conserved and are referred to as pribnow box (– 10 sequences) and – 35 sequence.

In eukaryotes, promotor region has three oligonucleotide sequences referred to as TATA box, GC box and CAAT box.

Q. What is pribnow box ?

Ans. Pribnow box is a sequence of six nucleotides which is rich in A and T. This is located 10 bp upstream and is also referred to as – 10 sequence.

Q. What is – 35 sequence ?

Ans. This is a sequence of 6 nucleotides located about 35 bp upstream the transcription start site.

Q. What is TATA box ?

Ans. TATA box is a sequence of nucleotides located about 25 bp upstream the transcription start site, in eukaryotes. It is a consensus sequence present in the promotor region.

TATA box binds with the TATA-biding protein (TBP) that is further associated with several other proteins, referred to as TBP-associated factors (TAF).

The whole complex of TBP and TAF is referred to as transcription factor. It facilitates the binding of RNA polymerase on the DNA template.

Q. What are the GC box and CAAT box ?

Ans. These are the nucleotide sequences that are located between – 110 bp and – 40 bp upstream. There are the binding sites for various transcription factors that affect polymerase functions and determine the frequency of transcription, in eukaryotes.

Q. Define enhancers ?

Ans. Enhancers are the gene-specific sequences, for the binding of transcription factors which, in turn, modulate binding of RNA polymerase to the promotor region, in eukaryotes. These can function in either orientation and are also referred to as promotor elements.These may be located several hundreds bp away from the promotor region and regulate the expression of gene by several folds.

Q. Define the action of RNA polymerase ? Discuss various types of RNA polymerases found in prokaryotes and eukaryotes ?

Ans. RNA polymerase synthesizes RNA using a DNA template. It does not need a primer.

Different types of RNA polymerases are found in prokaryotes and eukaryotes.

In eukaryotes, three types of RNA polymerases are found, each one of which is specific for the synthesis of a particular class of RNA. These are designated as RNA polymerase I, II and III.

RNA polymerase I transcribes mRNA genes and synthesizes pre-mRNAs and SnRNAs.

RNA polymerase III transcribes genes for tRNA while RNA polymerase II transcribes genes for rRNA.

Besides, there also exists mitochondrial RNA polymerase for the transcription of different types of RNAs with in the mitochondira.

Prokaryotes, on the other hand, contain a single RNA polymerase which is responsible for the synthesis of all types of RNAs. This is a multienzyme complex and consists of the core enzyme and a sigma factor. The core enzyme has four peptides referred to as α, α, β, and β′, *i.e.* 2α subunits, one β′ subunit and one β′ subunit.

The core enzyme, in association with the sigma factor forms holoenzyme.

Q. Describe the process of transcription in prokaryotes ? How does it differ in eukaryotes ?

Ans. In prokaryotes, synthesis of RNA is initiated with the binding of RNA polymerase holoenzyme to the DNA template. The enzyme recognizes the promotor and initiates transcription. Usually, a purine ribonucleotide is incorporated first. After holoenzyme initiates the process, sigma factor is released and the core enzyme continues the process. Topoisomerases are also important for elongation. After the desired gene has been transcribed, RNA polymerase recognizes the transcription-termination sequence and terminates the process.

The process of transcription in eukaryotes is more complex and requires several other proteins. Moreover, three types of RNAs are transcribed by three distinct RNA polymerases.

Q. Differentiate between the process of Rho-dependent and Rho-independent termination of transcription ?

Ans. Rho-dependent transcription termination requires the presence of a protein designated as Rho factor. It is a hexameric protein with RNA dependent ATPase activity. The nascent RNA wraps around the Rho factor which, in turn, results in the hydrolysis of ATP and causes the dissociation of RNA transcript from the DNA template.

Rho-independent termination : Transcription termination occurs after RNA polymerase has recognizesd the termination sequence. These are conserved concensus sequences of forty nucleotide basepairs containing GC rich hyphenated inverted repeats. As transcription proceeds through the hyphenated inverted repeat, the RNA transcript forms an intramolecular hairpin structure with GC nucleotides. This proceeds a sequence of several U residues in the chain. Formation of intramolecular hairpin results in the release of RNA transcript.

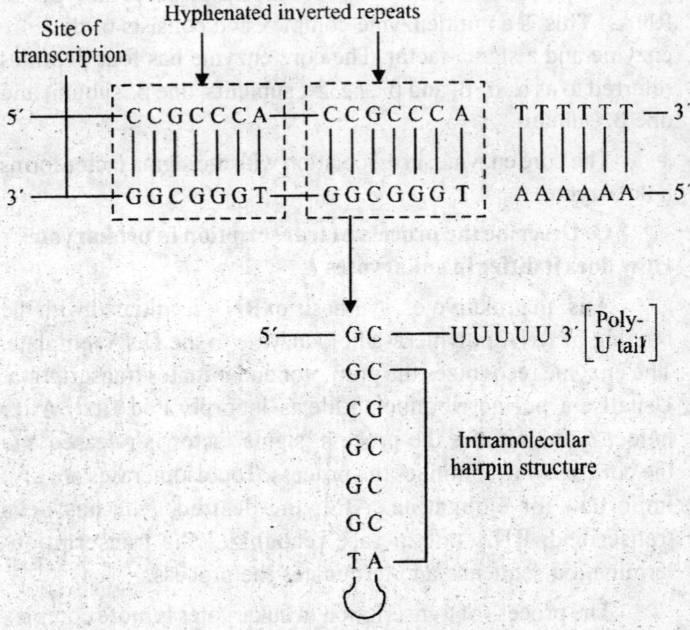

Primary RNA transcript

Q. Describe post-transcriptional modifications of RNA.

Ans. RNAs are processed in prokaryotes as well as eukaryotes.

mRNA : In eukaryotes, primary mRNA transcript, referred to as pre-mRNA is modified, as follows :

First, there is polyadenylation of pre-mRNA. Nearly 40 to 200 adenine nucleotides are added towards the 3′-end, by a soluble polymerase. This polyA tail helps in stablization of mRNA transcript and in its exit from the nucleus.

Primary mRNA also contains several coding as well as noncoding sequences which are referred to as exons and introns, respectively. Exons are the sequences which are expressed and retained in the mRNA. The noncoding sequences are not expressed and removed during post-transcriptional modification when pre-mRNA is changed to mRNA. The pre mRNA is processed by ribozymes which are small-nuclear ribonucleoproteins (snrnps). They bind to pre-mRNA and result in the splicing of introns and the joining of exons, leading to the formation of the functional mRNA.

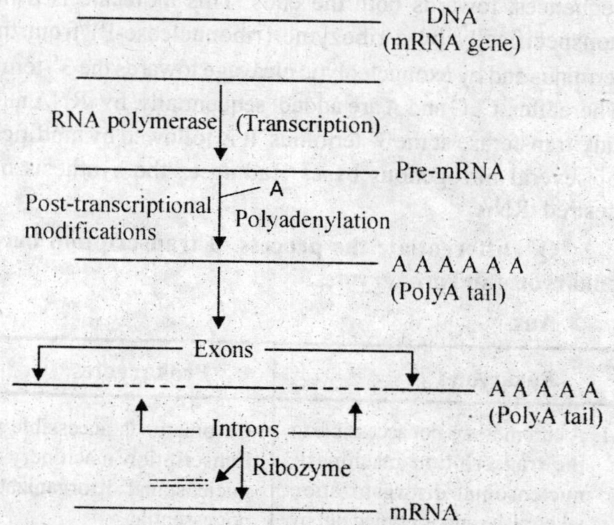

Incompletely processed mRNA present in the nucleus of a eukaryotic cell is referred to as heterogenenous nuclear RNA (hnRNA).

In prokaryotes mRNA is usually not modified

rRNA : Pre-rRNA molecules are synthesized as 45S RNAs, each one of the which has a sequence for three rRNA species and releases 28S, 5.8S and 18S rRNA. These rRNAs along with various proteins form components of the ribonucleoproteins, present in ribosomes.

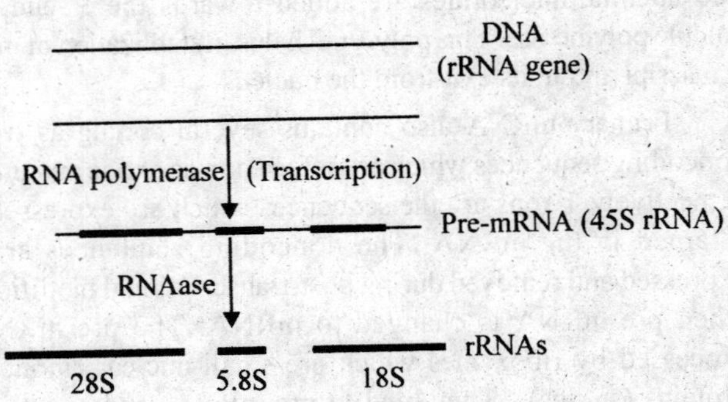

tRNA : Transcription of tRNA gene results in the formation of a primary transcript which has several extra nucleotide sequences, towards both the ends. This molecule is trimmed nonspecifically by a ribozyme (ribonuclease-P) from the 5′ terminus and by exonucleolytic cleavage towards the 3′-terminal. Thereafter C, C and A are added, sequentially, by tRNA nucleotidyltransferase at the 3′ terminus. It is followed by modification of several nitrogenous bases leading to the synthesis of the desired tRNA.

Q. Differentiate the process of transcription between eukaryotes and prokaryotes.

Ans.

Eukaryotes	Prokaryotes
1. A chromotin is not accessible to the transcription machinery, nucleosomal disorganization takes place and is carried out by DNase	Chromatin is accessible to the transcription machinery hence nucleosomal disorganization is not essential
2. Promotor region is very complex and requires the binding of several transcription factors, besides RNA polymerase	Promotor region is simpler and does not require transcription factor but the binding of RNA polymerase only
3. Transcription of three types of RNAs is carried out by three	Transcription of all three types of RNAs is carried out by a single

different RNA polymerases	RNA polymerase
4. Pre-mRNA transcript has several introns and is processed to mRNA	Pre-mRNA is not much different from mRNA
5. Post-transcriptional modifications of all three types of RNAs occurs.	Post-transcription modifications occur for tRNA and rRNA only.

Q. What is translation ?

Ans. The process of protein biosynthesis is referred to as translation.

It is the process of biochemical translation of genetic information from the four letter language of nucleotides (in nucleic acids), into the twenty letter language of amino acids (in proteins).

Information present in mRNA is translated from 5′ end to 3′ end, resulting in the production of a polypeptide which is synthesized from the aminoterminal to the carboxyterminal.

Q. What is genetic code ?

Ans. Genetic code is a dictionary that defines a correspondence between the sequence of four nucleotides in an mRNA to the sequence of amino acids in a polypeptide. Each code is a triplet, *i.e.* has three nucleotides.

Q. What is a codon ?

Ans. A codon is a sequence of three nucleotides, *i.e.* a triplet, present in mRNA.

Q. Why a codon is a triplet ?

Ans. As there are only 4 types of nucleotides (2 purines and 2 pyrimidines) to encode 20 amino acids that are commonly found in a protein, hence a triplet is essential. Thus $(4)^3$, *i.e.* 64, nucleotide triplets code for 20 amino acids.

If two nucleotides code for an amino acid, only 16 amino acids *i.e.* $(4)^2$, will be incorporated.

Q. Describe salient features of the genetic code.

Ans. Salient features of the genetic code are :

1. **Degeneracy :** There are more than one codons for several amino acids. This is referred to as degeneracy.

2. **Punctuation :** Message (nucleotide sequences) on mRNA is read continuously without any punctuation in the sequence

of codons until a nonsense codon is reached and that there is no punctuation in the genetic code.

3. **Nonoverlapping :** Sequence of nucleotides is translated in continuation, *i.e.* the next codon starts after the end of the previous one, without overlapping.

4. **Universality :** The genetic code is universal, *i.e.* same codons code for an amino acid in all the organism except with some exceptions within the mitochondria.

Q. What is initiation codon ?

Ans. Generally, methionine is the first amino acid that is incorporated during protein synthesis. As it is coded by AUG hence, AUG is referred to as the initiation codon.

Q. What are nonsense codons or chain terminating triplets ?

Ans. There are three codons (out of 64) which do not code for an amino acid, hence these are referred to as nonsense codons. These are also designated as stop codons, as protein synthesis terminates at this point. These are UAG, UAA and UGA.

Q. Describe the process of protein biosynthesis.

Ans. The process of protein biosynthesis involves activation of amino acids, and initiation, elongation and termination of protein synthesis.

Activation of amino acid.

The process of protein biosynthesis starts with the activation of amino acids. An amino acid binds with the specific tRNA. This reaction is catalyzed with the help of amino acid activating enzyme also referred to as aminoacyl tRNA synthetase.

Amino acid activation occurs in two steps.

1. In the first step, an amino acid forms a complex with the enzyme (E) and AMP (from ATP).

2. Subsequently, AMP and enzyme (E) are dissociated while amino acid is attached with the tRNA.

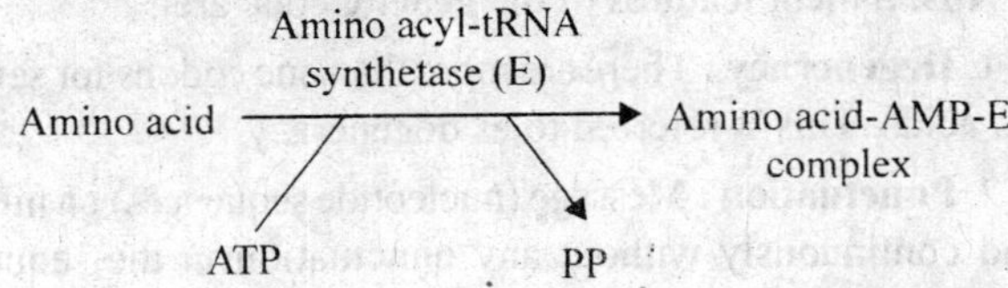

Amino acid-AMP-E
Amino acyl-tRNA
tRNA
AMP+Enzyme

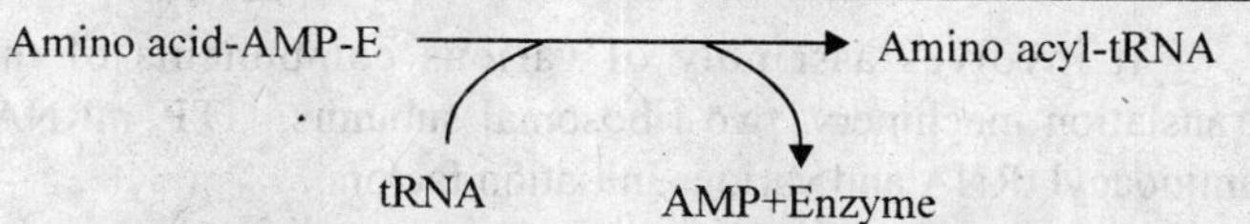

After the activation of the amino acid, the process of protein biosynthesis occurs in three steps.

Initiation

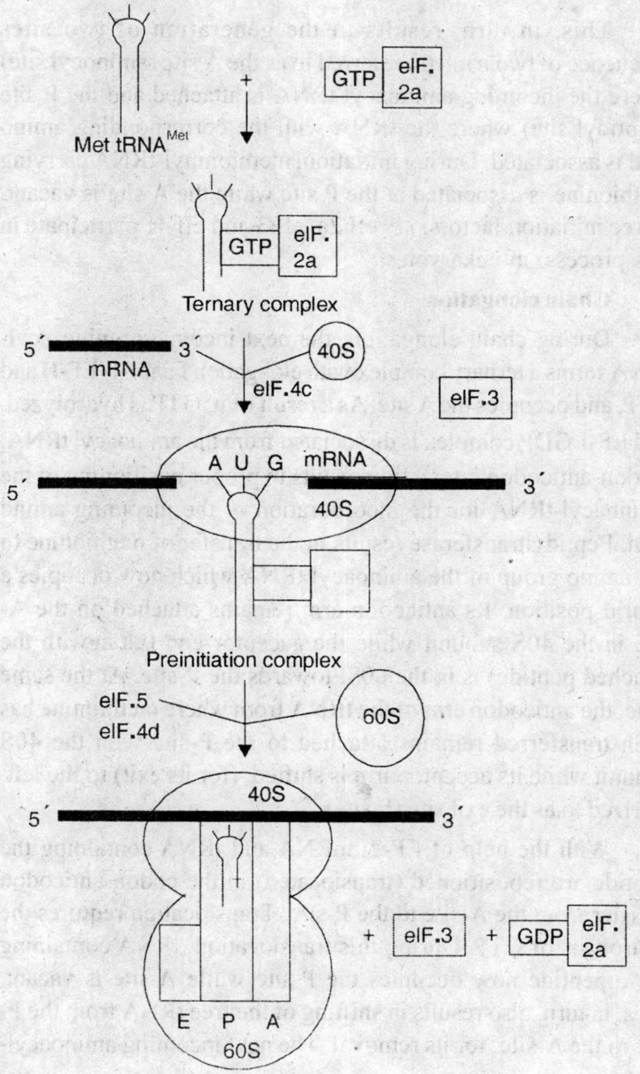

GTP
eIF.
2a
+
Met tRNA^{Met}
GTP
eIF.
2a
Ternary complex
5´
3´
mRNA
40S
eIF.4c
eIF.3
A U G
mRNA
5´
3´
40S
Preinitiation complex
eIF.5
eIF.4d
60S
40S
5´
3´
E P A
60S
+ eIF.3 + GDP eIF.
2a

It involves assembly of various components of the translation machinery, two ribosomal subunits, GTP, mRNA, aminoacyl tRNA and various initiation factors.

First, a pre-initiation complex which comprises of initiation factor, 40S ribosome, and methioninyl-tRNA (Met-$tRNA_{Met}$) interacts with mRNA and thereafter 60S ribosomal subunit is associated with it.

This, in turn, results in the generation of two sites (sequence of two triplets) referred to as the A site (aminocyl site) where the incoming aminoacyl tRNA is attached and the P site (peptidyl site) where the tRNA with the corresponding amino acid is associated. During initiation, methioninyl-tRNA carrying methionine is associated at the P site while the A site is vacant. Three initiation factors, *i.e.* eIF2a, eIF3 and eIF4c participate in this process, in eukaryotes.

Chain elongation.

During chain elongation, the next incoming amino acyl-tRNA forms a ternary complex with elongation factor-I (EF-I) and GTP, and occupies the A site. As a result of it, GTP is hydrolyzed, and EF-I.GDP complex is dissociated from the aminoacyl tRNA. Codon-anticodon interaction results in proper positioning of the aminoacyl-tRNA for the incorporation of the incoming amino acid. Peptidyltransferase results in the transfer of methionine to the amino group of the aminoacyl tRNA which now occupies a hybrid position. Its anticodon arm remains attached on the A-site in the 40S subunit while the acceptor end (alongwith the attached peptide) is in the 60S, towards the P-site. At the same time, the anticodon arm of the tRNA from where methionine has been transferred remains attached to the P-site with the 40S subunit while its acceptor arm is shifted, (for its exit) to the left, referred to as the exit site (E site).

With the help of EF-2, mRNA and tRNA containing the peptide are repositioned (translocated) in the codon-anticodon register from the A-site to the P-site. Translocation requires the hydrolysis of GTP. During this translocation, tRNA containing the dipeptide now occupies the P-site while A-site is vacant. This, in turn, also results in shifting of the free tRNA from the P-site to the A-site, for its removal. The next incoming aminoacyl-tRNA thereafter occupies the A-site and the process is repeated until the termination codon is reached.

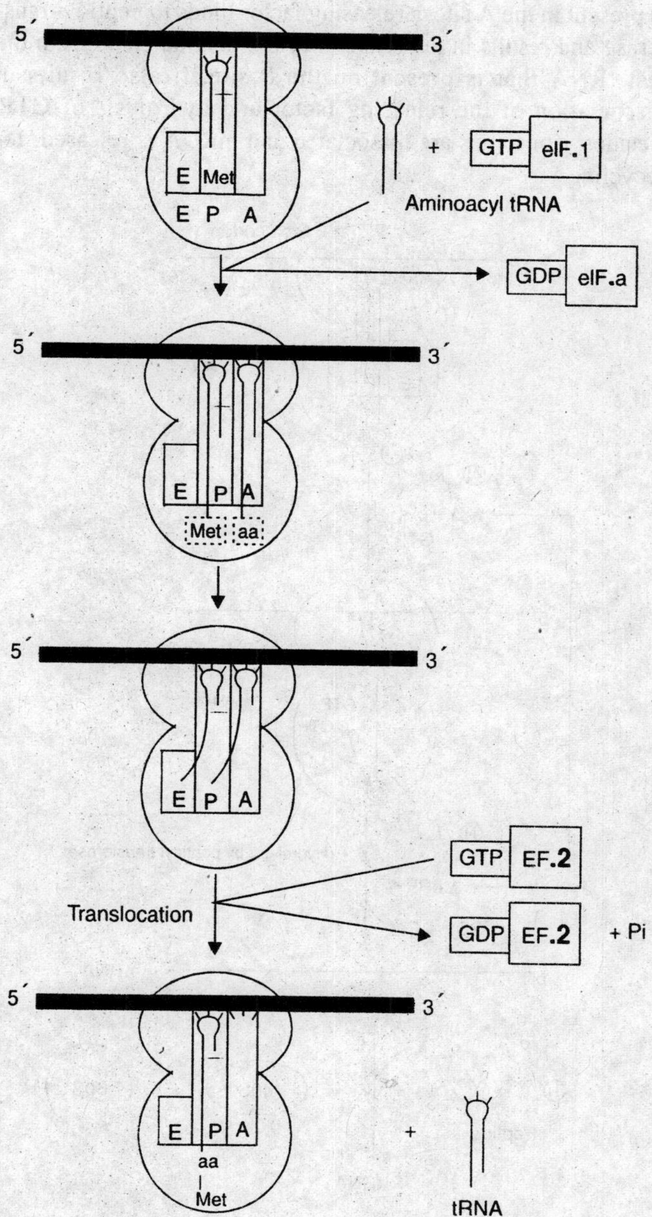

Termination

When one of three stop-codons, *i.e.* UAE, UAA or UGA, is present in the A site, a releasing factor binds to peptidyltransferase and results in the removal of the peptide chain from the last tRNA that is present on the P site. It also results in dissociation of the releasing factor and hydrolysis of GTP. Besides, ribosomes are dissociated and mRNA is released, for recycling.

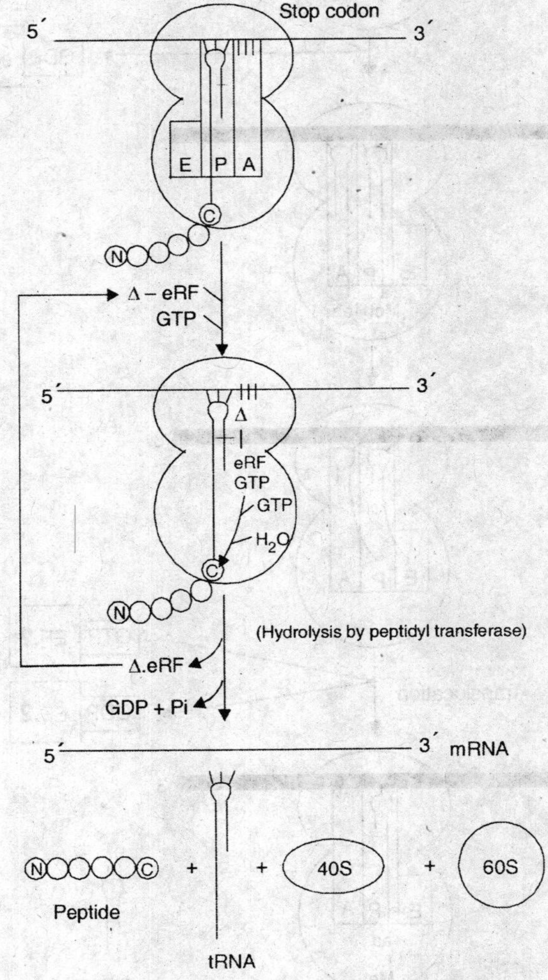

Q. Differentiate in protein biosynthesis between prokaryotes and eukaryotes.

Ans.

Prokaryotes	Eukaryotes
1. Ribosomal particles are 70S in size, comprising of 30S and 50S subunits	Ribosomal particles are 80S in size and comprise of 40S and 60S subunits
2. Free ribosomes are involved in the process of protein biosynthesis	Protein synthesis takes place on the free as well as bound ribosomes
3. Usually, mRNA is polycistronic	Usually, mRNA is monocistronic with polyA tail
4. First amino acid incorporated into the protein is N-formylmethionine	Methionine at the initiation point is not formylated
5. Three initiation factors participate in initiation of protein biosynthesis	Nine initiation factors participate in the process
6. Three elongation factors participate in chain elongation and are referred to as EF-TU, EF-TS and EF-G	Two elongation factors participate in chain elongation. These are referred to as eEF-1 and eEF-2
7. Three releasing factors, referred to as RF-1, RF-2 and RF-3, participate in chain termination.	Only one releasing factor (eRF) participates in chain termination

Q. What is Wobble hypothesis ?

Ans. Out of the three nucleotides in a triplet, the third base of the codon (*i.e.* the first base towards the 5′ end of the anticodon), encoding a particular amino acid, may vary from one codon to other. This is referred to as Wobble hypothesis. This position is degenerative and is less important as compared to other two nucleotides in the triplet.

Q. Define Shine-Dalgarno sequences ?

Ans. Shine-Dalgarno sequence is a purine-rich sequence of nearly ten nucleotides, found in prokaryotes. It is located upstream the initiation codon (AUG in mRNA).

This forms complementary basepair of pyrimidine-rich sequence of nucleotides on the 3′ end of 16S rRNA and facilitates the binding and positioning of mRNA on the 30S ribosomes.

Q. Explain inhibitors of protein biosynthesis ?

Ans. Several antibiotics and toxins are used therapeutically, to inhibit protein biosynthesis, both in prokaryotes as well as eukaryotes. Some of these inhibit the process of initiation of protein biosynthesis, others while interfere with chain elongation.

Inhibitor	Inhibitory in		Process inhibited
	Prokaryotes	Eukaryotes	
Streptomycin	30S subunit	-	Initiation
Tetracycline	30S subunit	40S subunit	Binding of aminoacyl tRNA
Puromycin and chloramphenicol	70S ribosome	80S ribosome	Peptide transfer
Erythromycin	50S subunit	-	Translocation
Diphtheria toxin	-	60S subunit	Translocation
Cyclohexamide	-	80S ribosome	Chain elongation

Q. Explain post-translational modifications of protein biosynthesis, with examples.

Ans. Modifications in a polypeptide chain, after its biosynthesis, are referred to as post-translational modifications.

A polypeptide may undergo post-translational modifications by partial proteolysis or covalent modifications.

Partial proteolysis :

A protein is synthesized as an inactive large molecule and is converted to small active protein, with the removal of a small peptide, *e.g.* conversion of preproinsulin to insulin, or pepsinogen to pepsin.

Covalent modifications :

In some of the proteins certain amino acids are modified, after their incorporation into the polypeptide chain, *e.g.* –

1. Hydroxylation of proline to hydroxyproline in collagen

2. γ-Carboxylation of glutamate to carboxyglutamate in certain blood clotting factors

3. Phosphorylation of serine and threonine in casein and several enzymes

4. Methylation of the ε-amino group of lysine in histones.

Q. What is an operon ?

Ans. An operon is referred to as a co-ordinated unit of gene expression, in prokaryotes. It consists of a set of clustered genes, which include –

1. Structural genes,

2. Regulatory gene, and

3. Control elements.

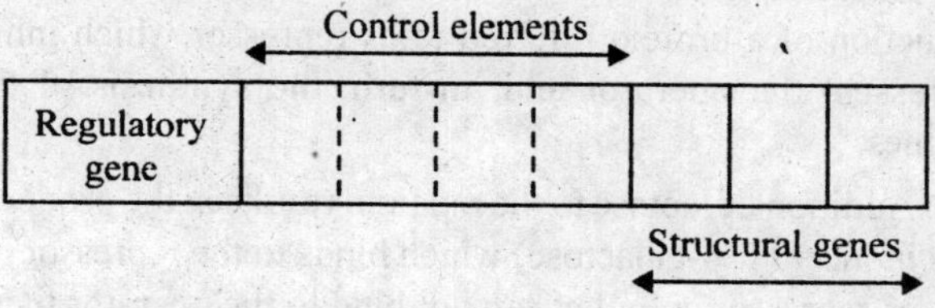

Q. What is induction of gene ?

Ans. Induction of gene is the process which results in increased expression of the gene, in response to certain signals which are referred to as inducers.

Q. What is gene repression ?

Ans. Repression is the process of turning-off of transcription, in response to some signal within a biosynthetic pathway. Such a signal is, generally, a product of the metabolic pathway and is referred to as a co-repressor.

Q. Describe Lac operon ?

Ans. Lac operon is a cluster of genes which regulate metabolism of lactose in bacteria such as E.coli. The Lac operon contains three adjacent structural genes which are referred to as Lac Z, Lac Y and Lac A. They code for β-galactosidase, permease and transacetylase, respectively. Besides three structural genes, Lac operon also contains a regulatory gene (Lac I), an operator (Lac O) and a promotor (Lac P).

Lac I encodes a protein referred to as repressor whereas

Lac O and Lac P are two control elements that are located in between Lac I and structural genes (the ZYA gene-cluster).

Bacteria, such as E.coli, usually depends upon glucose as a source of carbon as well as energy. However, when glucose is lacking, these bacteria can utilize lactose. Presence of lactose results in the induction of β-galactosidase that is essentially required for the metabolism of lactose, alongwith two other enzymes. This ability of bacteria to produce a series of proteins, only when needed for the utilization of a particular substrate, allows the bacteria to adopt to the environment by the regulation of Lac operon. This, in turn, explains need-based synthesis of proteins by the bacteria without continuously synthesizing them in large quantities when these are not required.

Expression of the regulatory gene (Lac I) results in the production of a protein referred to as repressor which inhibits (represses) the operator and, in turn, the synthesis of three enzymes.

Addition of lactose to the medium results in the production of an inducer (1, 6-galactose) which binds to the repressor. This inducer-repressor complex can not bind to the operator thereby resulting in transcription and translation of the gene. This is referred to as **Lac Repressor.**

Catabolite repression : Lac operon consists of three structural genes (Lac Z, Y and A), one regulatory gene (Lac I) and two control elements (Lac P and Lac O).

Lac O (operator) is a site for the binding of RNA polymerase whereas promotor (Lac P) has a site for the recognition of a protein that is referred as catabolite activator protein (CAP).

CAP is a DNA-binding protein and forms complex with cAMP. Thus, CAP is also referred to as cAMP-response protein (CRP). CAP-CRP complex, in turn, binds to the promotor.

Bacteria such as E.coli, though can utilize various sugars, however, it prefers glucose. Only when all the glucose has been utilized, it starts utilizing other sugars. This is due to the reason that when glucose is present in the medium, cAMP synthesis is inhibited and CAP-CRP complex is not formed. Hence, promotor does not promote the operator and thus the expression of the gene. On the other hand, when glucose is absent from the medium, cAMP levels are increased. This, in turn, results in the formation of CAP-CRP complex, which binds to the promotor for the expression of the structural genes.

Glucose a catabolite of lactose, thus, represses the induction of Lac operon.

Q. What is attenuation ?

Ans. Attenuation is a mechanism for the regulation of transcription through modulation of the nascent RNA secondary structure in prokaryotes. It regulates tryptophan operon transcription, depending upon the availability of tryptophan in the medium.

Q. Write note on tryptophan operon.

Ans. Tryptophan is an essential amino acid. However, if not present in required amounts in the medium, it can be synthesized in bacteria.

In E.coli, it is synthesized from chlorismic acid, in 5 steps by three enzymes, as two of these enzymes have two subunits each. Hence, for the synthesis of these five proteins, there is a cluster of 5 structural genes which are designated as trp A, trp B, trp C, trp D and trp E. Upstream the cluster of these structural genes, there is a promotor sequence (trp P) where transcription begins and an operator (trp O) for the binding of the repressor protein.

Tryptophan acts as a co-repressor and forms a complex with the repressor which, in turn, binds with operator, and inhibits the transcription of the structural genes and the synthesis of the required enzymes.

In addition, tryptophan operon also contains regulatory elements. One of these is referred to as a secondary promotor (trp P2) which is located with in the coding sequence of tryp D gene. This promotor is not regulated by the repressor and transcribes an incomplete mRNA for the expression of the last three genes of the operon, *i.e.* trp C, B and A. This second transcriptional central element is referred to as attenuation and is located upstream of trp E gene, in the leader sequence.

During trytophan deficiency, the entire polycistronic mRNA is synthesized. On the other hand, when tryptophan is present in the medium, it acts as a co-repressor and binds to the repressor and results in transcription of only the leader sequence.

Q. What is attenuator ?

Ans. Attenuator is a short open-reading frame, encoding a sequence of 14 amino acids that is referred to as a leader peptide. It has two adjacent tryptophan residues. This RNA sequence contains four complimentary segments which are referred to as segments 1, 2, 3 and 4. These can adopt several possible secondary structures. Out of these four segments, segments 3 and 4 alongwith the succeeding residues (GC rich hair-pin followed by several U residues) constitute a transcription terminator.

Q. Explain the process of regulation of gene expression in eukaryotes.

Ans. Gene regulation in eukaryotes is more complex than prokaryotes, because in eukaryotes most of the DNA is packed in inaccessible structure referred to as chromatin. Moreover, majority of the DNA is not transcribed and consists of repetitive sequences.

Several disruptive events take place which help in the conversion of the inactive chromatin into active chromatin, *e.g.* action of DNase I, acetylation of histone tails, demethylation of DNA sequences and several protein-protein interactions.

Gene expression in eukaryotes is also regulated at the level of translation.

Q. Differentiate in regulation of gene expression between prokaryotes and eukaryotes.

Ans. Regulation of gene expression in prokaryotes differs from eukaryotes, as follows :

Prokaryotes	Eukaryotes
1. Genome in prokaryotes is smaller in size and consists of a single circular chromosome (nearly 4.6×10^6 bases, in E.coli.)	The genome in quite large in eukaryotes, *e.g.* a human cell may contain 40,000 genes with nearly 3×10^9 bases.
2. Prokaryotes, in a particular species, have same type of cells containing a single chromosome.	Many eukaryotes have different types of cells, *e.g.* hepatocytes whose genes are highly expressed as compared to erythrocytes.

3. In prokaryotes, genes are organized in clusters in the form of operon.	Eukaryotic genes are not organized in operons but are spread widely across the genome.
4. Transcription and translation are coupled with each other.	As transcription and translation are uncoupled in eukaryotes, regulation of gene expression may occur either at the level of transcription or translation.

Q. What is recombinant DNA ?

Ans. Recombinant DNA is a molecule of DNA having fragments from different sources.

Q. What is recombinant DNA technology ?

Ans. Recombinant DNA technology, also called as genetic engineering or molecular cloning, refers to the processes that are used in the dissection of complex genomes into their well defined fragments, including complete analysis of nucleotide sequences and their functions.

Q. What are restriction endonucleases ?

Ans. Resctriction endonucleases are enzymes with restricted actions, *i.e.* they recognize and cleave specific DNA sequences which are symmetrical inverted repeats. These sequences are referred to as palindromes, such as

5′—G A A T T C—3′

3′—C T T A A G—5′

Restriction endonucleases are highly specific bacterial enzymes that make cut in each strand of the double-stranded DNA thereby generating two ends.

These are named, according to the source by using first letter of the genera and the first two letters of the species of the bacteria, *e.g.* EcoR1, Hae III, etc.

Most of these enzymes cleave both the strands of the DNA in a staggered manner and produce fragments with single-stranded regions, at their 5′ and 3′ ends. These ends are referred to as sticky ends (cohesive ends).

↓
5′—G AATTC—3′
3′—CTTAA G—5′
↑

EcoR1 ↓

5′—G AATTC—3′

+

3′—CTTAA G—5′

Some of these enzymes however, cut both the strands, symmetrically and produce fragments with base-paired ends. These ends are referred to as blunt ends (flush ends).

5′—GGCC—3′
3′—CCGG—5′

Hae III ↓

5′—GG3′ 5′CC—3′

+

3′—CC5′ 3′GG—5′

Q. What is a clone ?

Ans. A clone is a collection of the identical cells which contain a vector, *i.e.* carrying DNA of interest and arise from a common ancestor.

Q. What is a vector ?

Ans. A vector is a molecule of carrier DNA, to which a fragment of the foreign DNA (which is to be cloned) can be attached. Commonly used vectors include : plasmid, bacteriophage, cosmid, etc.

Q. What are plasmids ?

Ans. Plasmids are small circular DNA molecules containing few thousand basepairs. These are found in prokaryotes. Bacterial plasmids are most suitable for use as recombinant DNA vectors for the cloning of DNA segments of upto 10 kilobasepairs.

Q. What is a cosmid ?

Ans. Cosmid vectors are a cross between plasmid and bacteriophage. Cosmids contain an antibiotic resistant gene and can accommodate foreign DNA insert of upto 45 kilobasepairs.

Q. What is genomic library ?

Ans. Genomic library is a set of cloned DNA fragments, representing the entire genome of the organism. It thus refers to a collection of all fragments of the double-stranded DNA which are obtained by the digestion of total DNA of the organism with the restriction endonuclease and their subsequent ligation to an appropriate vector.

Q. What is a complementary DNA ?

Ans. Complementary DNA is the strand of DNA which is synthesized by the transcription of a functional mRNA with the use of RNA dependent-DNA polymerase (reverse transcriptase) in the presence of deoxyribonucleotides.

Q. What is cDNA library ?

Ans. The collection of the cloned cDNAs, synthesized from total mRNA in a given tissue or cell type is called cDNA library. cDNA library thus represents the population of mRNAs in a tissue.

Q, Describe the process of recombinant DNA technology with its applications.

Ans. Recombinant DNA technology involve several steps.

1. A fragment of the appropriate size of DNA of interest is generated, by using restriction endonuclease.

2. It is isolated and incorporated into a vector.

3. The vector with the DNA of interest is introduced into the cell for its replication.

4. Finally, cells containing the desired DNA are identified.

Recombinant DNA technology has various applications.

1. Diagnosis of prenatal defects : DNA isolated from cells of amniotic fluid of pregnant woman can be analysed for genetic defects in the fetus.

2. Medico-legal purposes : DNA from a small material available from the site of a crime such as a single hair, drop of

blood or sperm can be isolated, amplified and mapped for the identification of the culprit.

3. Gene therapy : For the selective introduction of a defective gene.

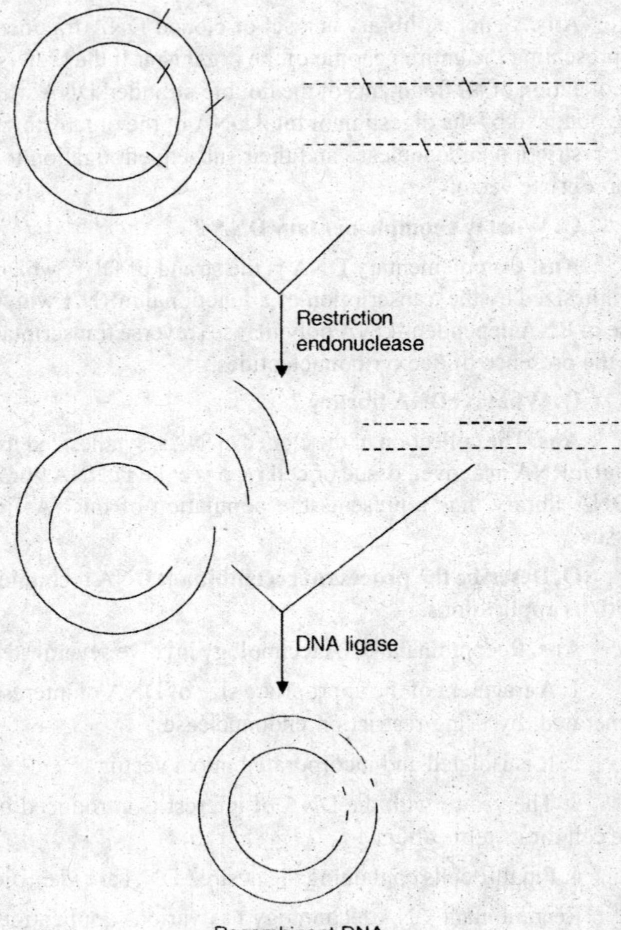

Recombinant DNA

Q. What is restriction fragment length polymorphism (RFLP) ?

Ans. RFLP refers to the process of cleaving of DNA into smaller fragments with restriction endonuclease to identify inherited differences in DNA sequences among the members of the same species.

It is useful in the diagnosis of those inherited disorders for which a molecular defect is not known, such as to define the clonal origin of a tumour.

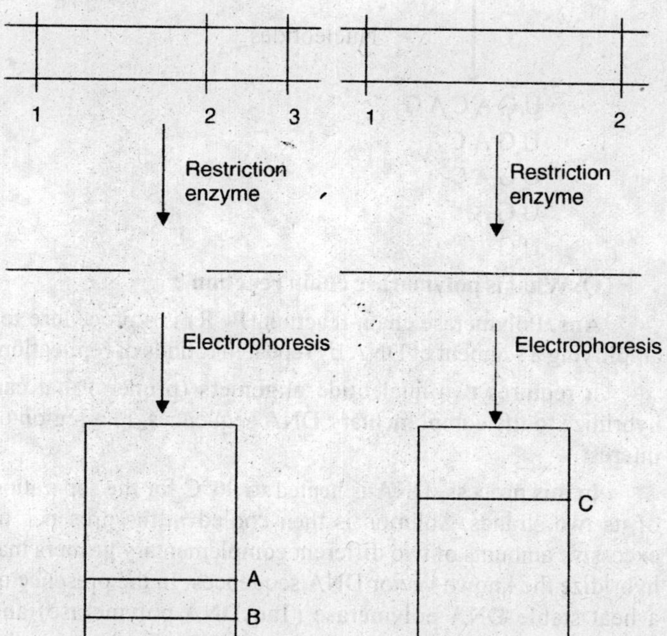

Q. What is DNA sequencing ?

Ans. DNA sequencing is the process of determining the sequence of nucleotides in DNA.

DNA is partially digested with phosphodiesterase which, in turn, results in a mixture of fragments of various lengths that can be separated.

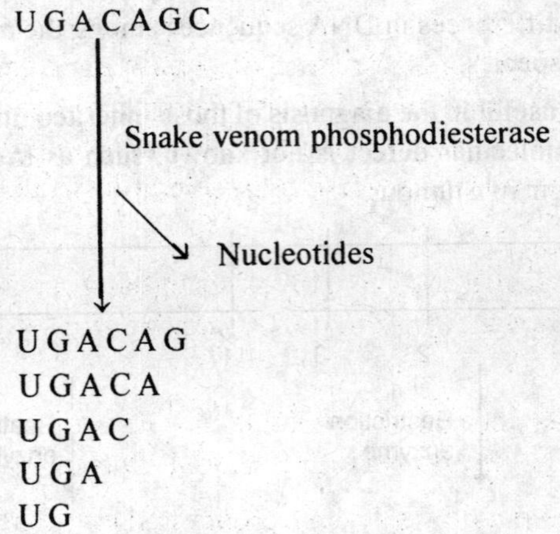

Q. What is polymerase chain reaction ?

Ans. Polymerase chain reaction (PCR) is a procedure for amplifying a segment of DNA by repeated rounds of replication.

It requires two nucleotide oligomers (primers) that can hybridize to the complimentary DNA sequences, in a region of interest.

In this process, DNA is heated to 90°C for the separation of its two strands. Solution is then cooled in the presence of excessive amounts of two different complementary primers that hybridize the known vector DNA sequences. In the presence of a heat stable DNA polymerase (Taq DNA polymerase) and deoxyribonucleotidetriphosphates (dNTP) there occur the synthesis of complementary strands. Multiple cycles of the process geometrically amplify the DNA.

Some of the applications of PCR include –

1. Diagnosis of infectious agent : PCR amplification of DNA sequences from the pathogen helps in the identification of infectious micro-organism, such as mycobacterium tuberculosis, HIV etc.

2. Forensic studies : Minute quantities of DNA such as from a drop of dried blood or sperm from the site of crime or clothes of the victim, can be amplified.

Q. What is Southern blot ?

Ans. Southern blotting is a technique for the identification of DNA base sequences after electrophoresis through its ability to hybridize with a complimentary single-stranded segment of the labelled DNA or RNA.

In this technique, DNA is isolated and cleaved into fragments by a restriction endonuclease. Various fragments are separated, according to their size, by electrophoresis. DNA is denatured and transferred on to the nitrocellulose membrane, using a radioactive probe. It was discovered by Edward Southern.

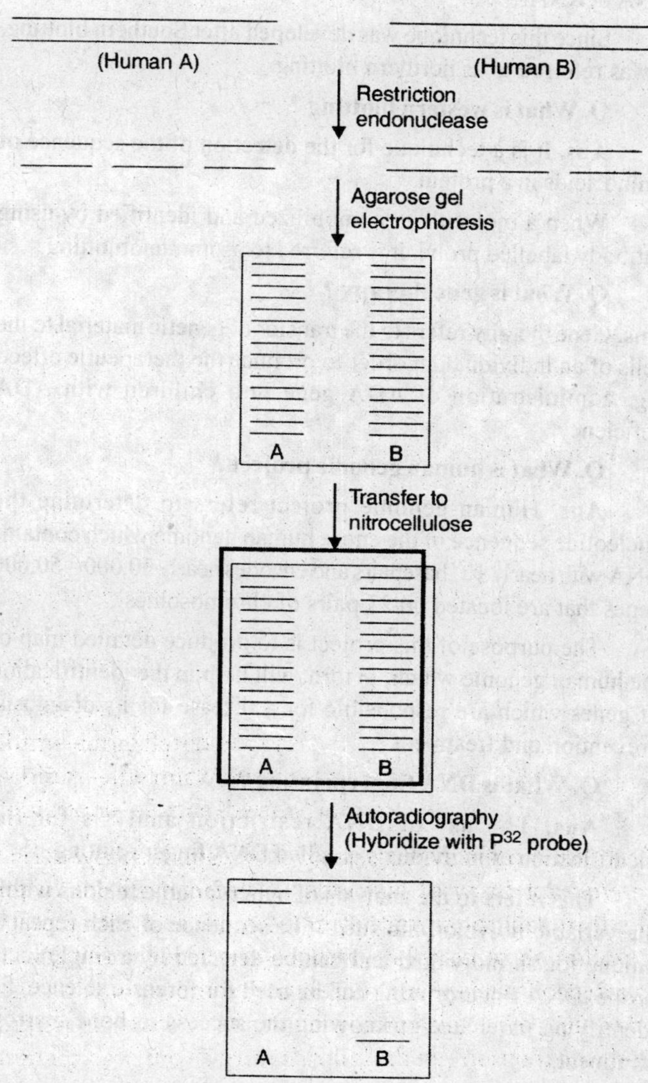

Q. What is northern blotting ?

Ans. Northern blotting is a technique for the identification of RNA, containing a particular sequence, through its ability to hybridize with a complementary single-stranded segment of the DNA or RNA.

Since this technique was developed after Southern blotting, it was referred to as northern blotting.

Q. What is western blotting ?

Ans. It is a technique for the detection of the sequence of amino acids in a protein.

When a molecule is immobilized and identified by using antibody labelled probe, it is referred to as immunoblotting.

Q. What is gene therapy ?

Ans. Gene therapy refers to the transfer of genetic material to the cells of an individual, in order to produce the therapeutic effect, *e.g.* administration of ADA gene into children with ADA deficiency.

Q. What is human genome project ?

Ans. Human genome project refers to determine the nucleotide sequence of the entire human genome which contains DNA with nearly 10^9 basepairs and encodes nearly 30,000 – 50,000 genes that are located on 23 pairs of chromosomes.

The purpose of this project is to produce detailed map of the human genome which, in turn, will help in the identification of genes which are responsible for a disease for its diagnosis, prevention and treatment.

Q. What is DNA fingerprinting ?

Ans. The use of DNA restriction analysis for the identification of individuals is called DNA finger printing.

This refers to the analysis of hypervariable regions within the variable polymorphic sites. The sequence of each repeat is unique for an individual and can be detected by a single locus probe. DNA fingerprinting can be used for forensic science, for identifying parentage or knowing the success of bone marrow treatment.

13

Acid Base Balance

Q. What is an acid?

Ans. An acid is a substance whose dissociation in water releases H^+ ions. Since H^+ does not contain a neutron and is essentially equivalent to a proton, therefore, an acid is also referred to as a proton donor.

$$HA \rightarrow H^+ + A^-$$

A strong acid such as HCl, gets dissociated completly whereas a weak acid, such as carbonic acid dissociates only partially. Most of the acids, produced in the body are weak acids.

Q. Define a base?

Ans. A base is a substance which release hydroxyl ion (OH^-) in an aqueous solution or accepts a proton. A strong base such as NaOH, dissociates completly whereas a weak base such as sodium bicarbonate, does not dissociate completly.

Q. What are amphoteric substances?

Ans. An amphoteric substance is a compound which can function, both as an acid as well as a base such as amino acids. At physiological pH, an amino acid can either donate a proton or accept it.

$$\underset{}{R-\underset{}{\overset{NH_2}{\overset{|}{CH}}}-COO^-} \xleftrightarrow[OH^-]{} R-\overset{NH_2}{\overset{|}{CH}}-COOH \xleftrightarrow[H^+]{} R-\overset{NH_3^+}{\overset{|}{CH}}-COOH$$

Q. What is a zwitterion ?

Ans. A zwitterion is an electrically neutral compound which exists, both as a cation as well as anion.

$$R-\underset{NH_3^+}{\underset{|}{CH}}-COO^-$$

Ampholytes exist as zwitterions at physiological pH.

Q. Define pH.

Ans. pH is defined as negative logarithm, to the base 10, of the molar H^+ concentration in a solution.

$$pH = -\log H^+$$

Low pH of a solution means it has high concentration of H^+.

Scale of pH varies from 0 to 14.

A solution is said to be acidic if its pH is below 7, neutral at pH 7 and alkaline if its pH is above 7.

Q. What is Henderson-Hasselbalch equation ?

Ans. Hendersson-Hasselbalch equation is a mathematical representation of the fundamental equilibrium equation which states that there is a direct relationship between pH and the ratio of the concentration of a conjugate base to the concentration of its acid.

Henderson-Hasselbalch equation is, as follows—

$$pH = pKa + \log \frac{(\text{Conjugate base})}{(\text{Conjugate acid})}$$

Q. What is a buffer ?

Ans. A buffer is a mixture of weak acid with its conjugate base or weak base and its conjugate acid.

Buffer minimises a change in pH with the slight addition of an acid or alkali. Effectiveness of a buffer depends upon the pKa of the buffer as well as absolute amount of the acid and the base.

Q. What are important buffers present in blood ?

Ans. Blood plasma is a mixed buffer system. It has three major buffers—

1. Bicarbonate buffer (HCO_3^- / CO_2),
2. Phosphate buffer $(HPO_4^{2-} / H_2PO_4^-)$, and
3. Protein buffer $(\text{Protein}^- / \text{H.Protein})$

In addition, RBCs also contain $Hb^- / H.Hb$ buffer.

Q. How is pH of the blood regulated ?

Ans. HCO_3^- / CO_2 buffer is most important in the regulation of pH, in the extracellular fluid since—

1. Two components of this buffer, *i.e.* HCO_3^- and CO_2 are present in large quantities, and

2. This buffer system operates as an open system, *i.e.* either of its component can be removed or added.

pH of the blood is regulated by two important tissues (lungs and kidneys) by acting on two components of this buffer.

Regulation by lungs : Lungs control partial pressure of CO_2 and thus minimise change in pH in the arterial blood. Normally, CO_2 is expired at the rate, at which it is produced and thus pCO_2 is maintained.

If blood becomes acidic, due to the addition of a fixed acid, hyperventilation occurs and CO_2 is expired at a greater rate, resulting in a decrease in pCO_2. On the other hand, if blood becomes alkaline, changes occur in the opposite direction.

Regulation by the kidneys : Kidneys regulate acid base balance by controlling the reabsorption of HCO^-_3 and secreting acids.

Q. What are different mechanisms of H^+ ion secretion?

Ans. H^+ ion secretion can occur by three mechanisms, *i.e.*—

1. By the reabsorption of the filtered HCO_3^-
2. By the excretion of titrable acids, and
3. By the excretion of ammonia.

Reabsorption of the filtered HCO_3^-

In the pulmonary capillary, CO_2 reacts with water and forms H_2CO_3 by a zinc containing enzyme carbonic anhydrase. This enzyme also catalyzes the dissociation of H_2CO_3 to $HCO_3^- + H^+$.

$$CO_2 + H_2O \longrightarrow H_2CO_3$$
$$H_2CO_3 \longrightarrow HCO_3^- + H^+$$

H^+ ions are actively secreted into the tubular fluid, in exchange for Na^+ and combine with HCO_3^- that has been filtered by glomeruli. This, in turn, leads to the formation of carbonic acid in the tubular fluid. Carbonic acid is dissociated into CO_2 and H_2O in the proximal convulated tubule. CO_2 is rapidly diffused through the cell membrane, combines with H_2O and forms carbonic acid. The H^+ are secreted into the lumen while HCO_3^- ions exit from the basolateral side of the cell along with Na^+.

The filtered HCO_3^- thus is reabsorbed indirectly by the secretion of H^+.

Blood	Tubule cell	Lumen
	$H_2O + CO_2 \rightleftharpoons H_2CO_3$	
	$\updownarrow$	HCO_3
$HCO_3 \leftarrow$	$HCO_3 + H^+ \rightarrow$	$H^+ \rightarrow H_2CO_3$
$Na^+ \leftarrow$	$Na^+ \leftarrow$	Na^+
		$\updownarrow$
		$H_2O + CO_2$

Excretion of titrable acids

The H^+ ions, that are secreted into the tubular fluid, titrate with phosphate buffer in the acidic direction, *i.e.* the basic form of the phosphate buffer (HPO_4^{2-}) is converted into its acidic form ($H_2PO_4^-$).

$$HPO_4^{2-} + H^+ \longrightarrow H_2PO_4^-$$

Excretion of titrable acids not only eliminate H^+ ions but also restores plasma HCO_3^- concentration since for each H^+ excreted as titrable acid, HCO_3^- is added to the blood.

Blood	Tubule cell	Lumen
	$H_2O + CO_2 \rightleftharpoons H_2CO_3$	
	$\updownarrow$	HPO_4^2
$HCO_3 \leftarrow$	$HCO_3^- + H^+ \rightarrow$	$H^+ \rightarrow H_2PO_4^-$
$Na^+ \leftarrow$	$Na^+ \leftarrow$	Na^+

Excretion of Ammonia

Ammonia is synthesized in the proximal tubule by the glutaminase reaction.

$$\text{Glutamine} \xrightarrow{\text{Glutaminase}} \text{Glutamic acid} + NH_3$$

NH_3 readly gets diffused through the cell membrane and combines with H^+ to form NH_4^+. These NH_4^+ ions are usually accompanied by some anion, such as Cl^-. Elimination of NH_4^+ ions in the urine contribute to net acid excretion as NH_4^+ is a major urinary acid. For each H^+ excreted in the urine as NH_4^+, HCO_3^- is added to the blood.

Blood	Tubule cell	Lumen
	$H_2O + CO_2 \rightleftharpoons H_2CO_3$	
	$\Updownarrow$	
$HCO_3 \leftarrow$	$HCO_3^- + H^+ \rightarrow$	$H^+ \rightarrow NH_4^+$
$Na^+ \leftarrow$	$Na^+ \leftarrow$	Na^+
Amino acids $\rightarrow$	Amino acids (Glutamine) $\rightarrow$	NH_3 Cl^-

Q. What is titrable acidity of urine ?

Ans. Amount of acid excreted is measured by the titration of urine to the pH of blood (7.4), and the quantity of the base so required is equivalent to the quantity of the acid excreted.This is referred to as titrable acidity of the urine.

The formation of titrable acidity accounts for nearly one-third to one-half of the normal daily acid excretion (by phosphate buffer) as most of the titrable acid, normally present in the urine, is inorganic phosphate.

Q. What is total acidity of the urine ?

Ans. Total acidity of the urine refers to the amount of total acids excreted. It is the sum of titrable acidity and acidity due to NH_4^+.

Q. What is acidosis ?

Ans. Acidosis refers to a decrease in blood pH below 7.35.

It may be a result of either excess of acid or deficiency of base.

Q. What is alkalosis ?

Ans. Alkalosis refers to an increase in blood pH above 7.45. It may be either due to the excess of a base or deficiency of an acid.

Q. What is respiratory acidosis ?

Ans. Respiratory acidosis refers to a fall in blood pH, as a result of the retention of CO_2. It may be due to hypoventilation such as in airways obstruction or a pulmonary disease which, in turn, results in increased CO_2 and a fall in pH. Kidney compensates for respiratory acidosis by increasing the retention of HCO_3.

Q. What is respiratory alkalosis ?

Ans. Respiratory alkalosis refers to the process of a rise in blood pH due to low pCO_2. This may be a result of hyperventilation which, in turn, causes a fall in arterial blood pCO_2 and a rise in blood pH, such as during anxiety, CNS injury involving respiratory centre, salicylate poisoning, etc., or at high altitude. Kidneys compensate for such a situation by excreting more HCO_3^-.

Q. What is metabolic acidosis ?

Ans. Metabolic acidosis refers to loss of HCO_3^-. It may also be due to the accumulation of nonvolatile acids such as lactic acid. Metabolic acidosis may occur during renal failure, uncontrolled diabetes mellitus, diarrhea or after severe excercise. Compensatory mechanism involves hyperventilation and finally increased H^+ excretion by the renal tubule.

Q. What is metabolic alkalosis ?

Ans. It is characterized by the intake of excess alkali or loss of fixed acids other than H_2CO_3. Metabolic Alkalosis may occur due to the ingestion of the excessive amount of sodium bicarbonate. Immediate compensation is by hypoventilation which is followed by increased renal excretion of HCO_3^-.

Q. List commonly measured parameters to know acid base status of an individual ?

Ans. Acid base status of an individual can be assessed by measurements of blood pH, pCO_2 and HCO_3^-. Their normal levels in the arterial blood are :

pH = 7.35 to 7.45

pCO_2 = 35 to 45 mm of Hg

HCO_3^- = 22 to 26 meq/L

Besides, base excess and anion gap can be calculated.

Q. What is anion gap ?

Ans. Anion gap is a difference between the sum of cations ($Na^+ + K^+$) from the sum of anions ($Cl^- + HCO_3^-$), *i.e.* it refers to plasma anions other than Cl^- and HCO_3^-, which are not routinely measured.

Anion gap = $[(Na^+ + K^+) - (Cl^- + HCO_3^-)]$

Normal value for anion gap is 12 – 16 meq/L.

14

Hormones

Q. What are hormones ?

Ans. Hormones are referred to as chemical messengers which transmit signals to generate alterations at the cellular level.

Q. What are Endocrine hormones ?

Ans. An endocrine hormone refers to a substance that is synthesized and secreted in one tissue and travel, through circulation to a long distance, to reach the target cell which expresses receptors for the same.

The tissue or gland secreting such substances is referred to as endocrine.

Q. What are paracrine hormones ?

Ans. Paracrine hormones refer to chemical messengers which are secreted by one tissue and travel relatively smaller distance to interact with their receptors. Such a substance mainly affect those cells which are adjoining the tissue that produces it.

Q. What are autocrine hormones ?

Ans. Autocrine hormones are the group of substances which are produced by the cells that also function as the target for it, *i.e.* they exhibit their action within the cell of origin.

Q. What are the important physiological functions of hormones?

Ans. Hormones perform several functions:

(*i*) The most important function of an endocrine hormone is the maintenance of chemical composition of the extracellular as well as intracellular fluid, *e.g.* maintenance of blood glucose by insulin, or Ca and P by PTH, etc.

(*ii*) Hormones are also important for the maintenance of physiological functions and growth of the organism, *e.g.* growth hormone.

(*iii*) Hormones also help in the production, use and storage of energy, *e.g.* thyroxine.

(*iv*) Hormones also help in the regulation of reproductive functions, *e.g.* hormones of the pituitary.

Q. Explain mechanisms of hormone action ?

Ans. Hormones act on their target cells either directly or through some other substance referred to as a secondary messenger. Some hormones function to regulate the release of other hormones from different cells, *e.g.* hormones released from the hypothalamus regulate secretion of the pituitary.

Q. What is hormone cascade system ?

Ans. Several hormones, particularly in higher organisms exhibit their action in series through various components of the system. Such a system is called hormone cascade system, *e.g.* a signal (stimulus) such as an electrical impulse or a chemical signal, originated either externally or internally, stimulates secretion of the chemical messenger from the hypothalamus (releasing/ inhibitory factors) which in turn regulate the secretion of hormones of the pituitary (trophic hormones). The signal is finally terminated with the release of the ultimate hormone by the target gland.

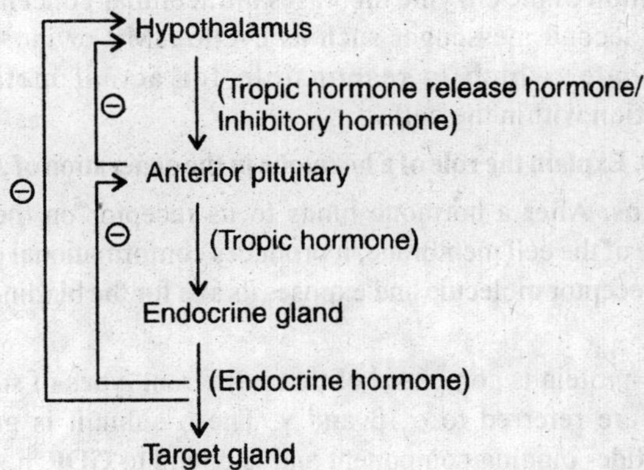

The production/secretion of the hormone is regulated through a series of negative (feed back) routes, *e.g.* hormones of a target gland may regulate the release of trophic hormones of the pituitary as well as hypothalamic factors.

Q. What is a hormone receptor ?

Ans. Hormone receptors are proteins that differ by their specificities for the ligand (hormone) as well as location within the cell. A receptor binds to its hormone and forms hormone-receptor complex which undergoes conformational changes and allows its interaction with a transducing protein on the cell membrane or acts intracellularly.

Q. Explain the mechanism of action of hormones acting through intracellular receptors ?

Ans. Several hormones gain access to the cytoplasm, by passive diffusion and bind to specific receptor protein forming hormone-receptor complex either with in the cytoplasm or in the nucleus. This, in turn, binds to specific regulatory sequences with in the DNA and regulates transcription. Such regulatory sequences are called hormone-regulatory elements.

Q. How do membrane-bound receptors function ?

Ans. A hormone binds to its receptor which is present in the outer plasma membrane of its target cell.

A hormone-receptor complex causes the activation of either adenylate cyclase or phospholipase C, both of which are located on the inner side of the plasma membrane. Activation of the enzyme increases intracellular concentration of the second messenger such as cyclic AMP or inositoltri-phosphate, which is responsible for actual metabolic regulation within the cell.

Q. Explain the role of a hormone in the generation of AMP ?

Ans. After a hormone binds to its receptor on the outer surface of the cell membrane, it produces conformational change in the receptor molecule and exposes its site for the binding of G-protein.

G-protein is composed of three different types of subunits which are referred to α, β and γ. The α-subunit is guanine nucleotide- binding component and is bound to GDP.

The activated receptor interacts with the $\beta - \gamma$ fraction of the G-protein and activates it. Activation of the receptor also causes the release of GTP from the α-subunit and its separation from the $\beta - \gamma$ subunits.

Free G-α, in turn, binds to adenylate cyclase and stimulates the synthesis of cyclic AMP. During this process, GTP is hydrolysed by the GTPase activity of the α-subunit, which thereafter returns to its original conformation, *i.e.* gets reassociated with the $\beta - \gamma$ subunits of the G-protein.

The G-protein can either be stimulatory (Gs) which mediate adenylate cyclase stimulation, or inhibitory (Gi) which mediates adenylate cyclase inhibition.

Q. 12. Explain the role of cAMP as second messenger ?

Ans. cAMP activates protein kinase A which is present as a tetramer. It has two regulatory (R) subunits and two catalytic (C) subunits, and requires four molecules of cAMP for its activation. Association of four molecules of cAMP with the two regulatory subunits results in the release of the catalytic subunits (active protein kinase).

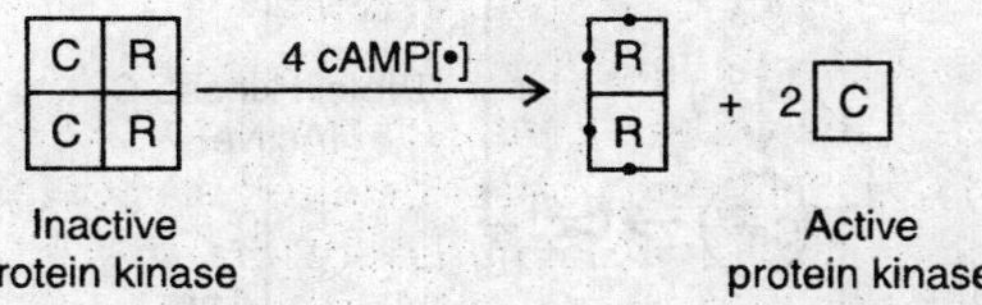

The active enzyme, in turn, phosphorylates target proteins to produce a cellular effect.

Q. Explain the mechanism of action of phosphatidyl inositol system as second messenger ?

Ans. Binding of some of hormones with the specific cell membrane receptor results in the stimulation of phospholipase C. The enzyme, in turn, catalyzes the hydrolysis of phosphatidylinositol-4, 5-bisphosphate (PIP_2) and results in the formation of diacylglycerol (DAG) and inositol -1, 4, 5-triphosphate (IP_3), both of which act as second messengers.

IP_3 diffuses into the cytosol and binds to its receptor on the membrane of the endoplasmic reticulum, and causes the release of free calcium to the cytosol. Both, increased intracellular calcium as well as DAG activate protein kinase C.

Activated protein kinase C, in turn, causes phosphorylation of specific proteins and exhibit various metabolic functions.

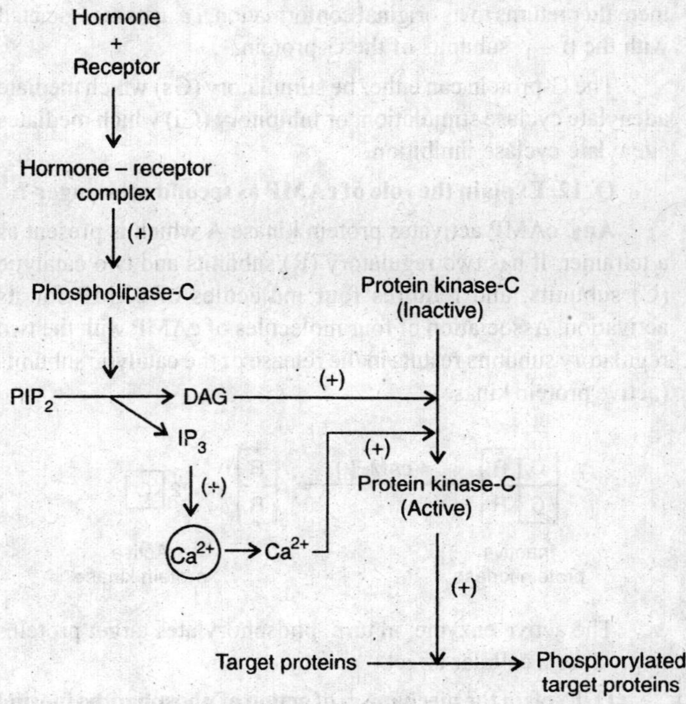

Q. Classify hormones with respect to their chemical nature?

Ans. According to their chemical nature hormones are classified into three groups:

(*i*) Peptide hormones, *e.g.* growth hormone

(*ii*) Steroid hormones, *e.g.* cortisol

(*iii*) Amino acid derivatives, *e.g.* epinephrine.

Q. 15. Name hormones released from the hypothalamus ?

Ans. Hypothalamus produces several hormones which control secretion of hormones of the pituitary gland. These are referred to as release hormones or release inhibitory hormones. These include :

Thyrotropin release hormone (TRH),

Corticotropin release hormone (CRH),

Luteinizing hormone release hormone (LHRH),

Growth hormone release hormone (GHRH),

Growth hormone release inhibitory hormone (GHIH),

Prolactin releasing factor (PRF), and

Prolactin release inhibitory factor (PIF).

Q. 16. Name hormones secreted by the anterior pituitary ?

Ans. Anterior pituitary (adenohypophysis) produces following hormones:

Growth hormone (GH),

Thyroid stimulating hormone (TSH),

Follicle stimulating hormone (FSH),

Luteinizing hormone (LH),

Prolactin,

Adrenocorticotropin,

β-Lipotropin, and

α-Melanocyte stimulating hormone (MSH).

TSH, FSH and LH are glycoprotein whereas prolactin, growth hormone and ACTH are protein in nature. The remaining hormones are synthesized as a single polypeptide referred to as pro-opiomelanocortin (POMC).

Q. What is pro-opiomelanocortin ?

Ans. Pro-opiomelanocortin (POMC) is a polypeptide which is a product of a single gene but is a precursor of several hormones. In the corticotropic cells of the anterior pituitary, it is cleaved to release ACTH and β-lipotropin. In the paras intermedia, these polypeptides are further cleaved to release α-MSH, corticotropin like interme-diary peptide (CLIP), γ-lipotropin and β-endorphin. First thirteen amino acids of ACTH contain α-MSH sequence.

15

Immunology

Q. What is an antigen ?

Ans. An antigen, also called an immunogen, is a foreign macromolecule that triggers immune response and induces antibody production. An immunogen can be specifically recognised by the adaptive elements of the immune system, T-cells and B-cells.

Q. What is an epitope ?

Ans. An epitope is referred to a part of the antigen, to which antibody binds. An antigen can have several epitopes.

Q. What is a hapten ?

Ans. Hapten is a small molecule which can act as an epitope but is not capable of exhibiting an antibody response of its own. A hapten can induce synthesis of antibodies only when it is covalently attached to some large molecule.

Q. What is an allergen ?

Ans. An allergen is an antigen that gives rise to immediate hypersensitivity, *e.g.* pollen, dust, etc. An allergen causes IgE mediated hypersensitivity.

Q. What is hypersensitivity ?

Ans. Some immunogens, such as allergens, exhibit excessive immune reaction (allergic response) which is mediated by the binding of IgE antibody to the mast cell. This is referred to as hypersensitivity.

Q. Explain various types of hypersensitivity reactions.

Ans. There are four types of hypersensitivity reactions, *i.e.*–

1. **Type I or immediate hypersensitivity :** It is characterized by the production of IgE antibodies against foreign proteins which are commonly present in the environment, *e.g.* pollens. These antibodies, specifically, bind to high affinity receptors on mast cells and basophils which contain histamine, *e.g.* asthma, hay fever, etc.

2. **Type II hypersensitivity :** It refers to antibody mediated hypersensitivity. These reactions are caused by IgG or IgM antibodies against cell surface and extracellular matrix antigens. Antibodies trigger cytotoxic reactions, and damage cells and tissues by activating complement, *e.g.* autoimmune hemolytic anemia.

3. **Type III hypersensitivity :** It refers to an immune complex disease. Accumulation of the immune complex triggers either a complement or cell mediated local reaction, *e.g.* systemic lupus erythromatosis.

4. **Type IV hypersensitivity :** It is characterized by cell mediated reactions. In these reactions, specific T cells are the primary effector cells, *e.g.* graft rejection.

Q. Define complement system and its functions ?

Ans. Complement system is one of the major defence systems of body. It consists of nearly 30 – 40 proteins which interact with each other and with other elements of the immune system.

Various components of the complement system are activated sequentially. Complement system controls inflammatory reactions and chemotaxis, clearance of immune complexes, cellular activation and antimicrobial defects. An important role of the complement system is to generate opsonins, proteins that stimulate phagocytosis by neutrophils and macrophages.

Q. What are immunoglobulins ?

Ans. Immunoglobulins are a group of glycoproteins present in the serum and tissue fluids of all mammals, and are also called antibodies. These are synthesized and secreted by the plasma cells, in response to invasion by an antigen.

Q. Describe various structural features of an immunoglobulin molecule ?

Ans. Structural features of an immunoglobulin molecules are shown below –

1. Immunoglobulin molecule is a 'Y' shape multi protein complex, structurally exhibiting two arms and a stem.

2. It has four polypeptide chains, comprising of two small subunits called L-chains and two large subunits referred to as H-chains. These are linked together by disulphide linkages.

3. Each subunit has a variable region towards the N-terminal-end and a constant region towards the C-terminal end.

4. Two arms have antibodies binding sites and are referred to as F_{ab} fragments (antibody binding fragments) whereas the stem is referred to as the F_c fragment (crystallizable fragment)

5. Its F_{ab} fragments are linked with the F_c fragment through a flexible region referred to as hinge region.

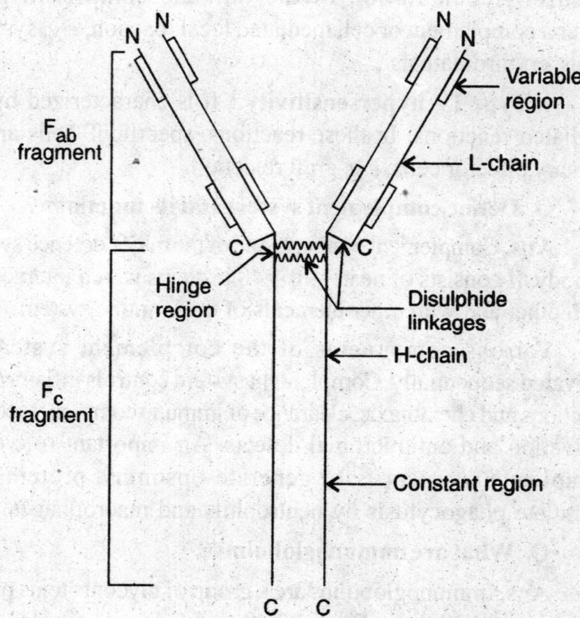

Q. What are different light chains in immunoglobulins ?

Ans. Light chains (L-chains) are small polypeptides (about 25 kDa each) containing nearly 220 amino acids. L–chains may be kappa (κ) chain or a lambda (λ). Either of the two of L-chains may be found in an immunoglobulin *i.e.* an immunoglobulin molecule may contain either 2κ chains or 2λ chains but not a mixture of one each.

Q. What are different heavy chains present in immunoglobulins ?

Ans. Heavy chains (H-chains) are large polypeptide chains containing 440 to 450 amino acids. Their molecular mass may vary from 50 to 75 kDa.

Five different types of H-chains, specific to each class of an immunoglobulin, are found, as shown below :

Immunoglobulin type	H-chain found
IgA	α (alpha)
IgG	γ (gamma)
IgM	μ (mue)
IgD	δ (delta)
IgE	ε (epsilon)

Q. What are variable and constant regions in an immunoglobulin chain ?

Ans. Nearly half of the N-terminal L-chain and one quarter of the H-chain have variable regions. These are heterogenous to the extent that no two variable regions from different individuals have identical amino acid sequences. These variable regions determine the specificity of an antigen to the antibody. These are designated as V_L and V_H regions in the light chain and the heavy chain, respectively.

The remaining half of C-terminal of L-chain and three-quarters of C–terminal segment of the H-chain have amino acid sequences which are homologous with the other L or H chains of the same class of immunoglobulins. These segments are called as constant regions. These regions in the two types of chains are referred to as C_L and C_H regions, respectively.

In the L-chain there is one C_H region whereas there are three C_H regions in H chain. These are referred to as C_H1, C_H2, C_H3 , respectively. Each of these C_H regions has nearly 110 amino acids. These are homologous to each other as well as to the C_L region. Each of these homologous sequences contains an intrachain disulfide linkage.

Q. What is the significance of C_H regions ?

Ans. 1. C_H regions determine class of an antibody, *i.e.* differences in the sequence of C_H region are characteristic of each class of immunoglobulin.

2. This is the site for the binding of a complement protein.

3. Through the C_H region, an immunoglobulin crosses the placental membrane.

4. Variable amounts of carbohydrates are attached to the immunoglobulin molecule within the C_H2 region of the H-chain.

5. C_H region is also linked through the joining peptide (J-chain) during polymerization of the immunoglobulin molecule, such as in IgA and IgM.

Q. What are hypervariable sequences ?

Ans. Hypervariable sequences are few short sequences of amino acids present within the variable regions. These are responsible for the recognition and binding of an antigen. In each hypervariable sequence, only a small peptide of 5–10 amino acids contributes to an antigen-binding site. These sites since are complementary to the topology of an antigen, are also designated as complementarity determining regions.

There are 3 hypervariable sequences in a V_L region and 3-4 in each of the V_H region, depending upon the type of an immunoglobulin.

Q. What are frame-work regions ?

Ans. Amino acid segments which separate different hypervariable sequences form each other, with in a variable region, are referred to as frame-work regions.

Q. What is the action of papain on an immunoglobulin molecule ?

Ans. Papain is a proteolytic enzyme which hydrolyzes the immunoglobulin molecule. It cleaves an immunoglobulin from the hinge region, forming three fragments. Two of these fragments are obtained from two arms of the immunoglobulin structure while the third fragment is obtained from the stem.

Two fragments, which are obtained from two arms are referred to as F_{ab} fragments, *i.e.* antigen binding sites whereas the other fragment is called as Fc fragment since it can be easily crystallized.

Q. What is a hinge region ?

Ans. Hinge region is a flexible region connecting two Fab fragments of the immunoglobulin molecule with the Fc fragment.

It is present between C_H1 and C_H2 , and allows two Fab fragments to move independently, to bind two antigens that may be present at a distance. Hinge region contains one or more cysteine residues which provide interchain disulphide bridges. This region is sensitive to the action of papain.

Q. Classify immunoglobulins.

Ans. Immunoglobulins are classified into 5 types, each one of which is designated with respect to the type of its heavy chain. These are referred to as IgG, IgA, IgM, IgD and IgE.

IgG : It is a major immunoglobulin produced by the plasma cells. It is the most important class of immunoglobulin, distributed in the intravascular as well as extravascular pool. It comprises of 70–75% of total serum immunoglobulin. IgG is produced in response to most bacteria and viruses. IgG type of antibodies bind to macrophages and neutrophils and promote phagocytosis.

IgG has four subclasses referred to as IgG_1, IgG_2 , IgG_3 and IgG_4. They differ from each other with in the hinge region. In human beings, all four can cross placenta and confer high degree of passive immunity to the newborn.

IgA : It accounts for 15–20% of total serum immunoglobulins. IgA type of antibodies may be present either as monomers or dimers. In human beings, more than 80% of IgA occurs as monomer.

IgA has two subclasses, *i.e.* IgA_1 and IgA_2. In human serum, IgA_1 is predominant. It is present in secretions such as nasal secretions, tear, saliva and milk.

Immunoglobulins which are present in secretions are also referred to as secretory immunoglobulins (sIgA).

sIgA, exists mainly, in the dimeric form, in association with a small protein chain, called secretory component. sIgA provides defence against invading viruses and bacteria.

IgM : It accounts for 5–8% of total circulatory immunoglobulins. These antibodies are present in serum as a pentamer where five molecules are arranged around a chain, referred to as the joining chain (J-chain). IgM also contains large amount of oligosaccharides that is associated with the μ-chain.

It is the first immunoglobulin which is synthesized after the introduction of an antigen in the host. It is also the only immunoglobulin that a neonate normally synthesizes. Hence, it is also called as the first line of defence. It is mainly present in intravascular compartment.

IgD : IgD is present in small amounts (<1% of total serum immunoglobulins). It is a major component of the surface membrane of several B-cells. Its primary function is not yet known.

IgE : It is present in minimum concentration and is found on the surface membrane of basophils and mast cells. It protects the host against parasites and is implicated in allergic reactions.

Q. What do you understand by antibody diversity ?

Ans. Antibody diversity refers to the potential of immune system, to produce large number of antibodies each one of which is specific for a type of antigen. This is achieved by gene recombination.

Q. How does immunoglobulin gene recombination take place ?

Ans. During the synthesis of immunoglobulins, both light and heavy chains are synthesized separately.

A light chain is a product of the cluster of 3 structural genes which are referred to as V-gene (variable region gene), J-gene (joining region gene) and C-gene (constant region gene). In addition, gene cluster for heavy chain also contains a D-gene segment (diversity gene).

Gene recombination for light chain and heavy chain thus occurs independently, resulting in the synthesis of IgM. Class switching thereafter results in the production of other types of immunoglobulins.

Q. Explain the process of gene recombination for the synthesis of a light chain.

Ans. Two separate segments of DNA code for the constant and the variable regions of a light chain.

The κ light chain is encoded by four exons which occur in clusters, *i. e.* the—

1. **Leader Segment (L_K) :** This segment encodes 17–20 amino acids signal peptide which, in turn, directs the newly

synthesized κ chains to the endoplasmic reticulum.

2. **Variable region segment (V_K) :** This segment encodes first 95 amino acids of the variable region of the κ chain.

3. **Joining chain segment (J_K) :** This segment encodes remaining 13–15 amino acids of the variable region.

4. **Constant region segment (C_K) :** This gene segment encodes for the constant region of kappa chain.

The kappa chain gene family consists of about 150 L_K and V_K gene segments. Downstream to the sequence there are five segments of J_K and a single segment of C_K. Various segments are separated from each other by introns. The Assembly of κ chain occurs within the B–cells by gene rearrangemet. During re-arrangement, a single $L_K - V_K$ unit is linked to a J_K segment with the deletion of the intervening sequences. Subsequently, modified gene is transcribed and spliced to form L – V – J unit. The leader sequence is cleaved and splicing of mRNA results in the synthesis of V–J combination.

The V–J segment is now linked to C_K segment which is translated into a protein.

Similar to gene recombination for the κ chain there occurs gene recombination for the λ chain. The λ gene locus, in human gene, contains a set of V-gene, 7 C-genes and one J-gene. Following rearrangement of V or J genes and the gene for C-region, DNA is transcribed into mRNA and finally translated into λ chain.

Q. How does gene recombination for heavy chain takes place ?

Ans. In human beings, heavy chain gene family consists of a cluster of nearly 250 sequential pairs of $L_H - V_H$ segments followed by 10 D segments, 6 J_H segments and 8 C_H segments. During lymphocyte differentiation, $L_H - V_H$ unit is joined to a D-segment and 2 J_H segments. During this process, D-segment is flanked by short-stretches of random sequences called N–regions. In the B-cell, transcription and splicing joins the $L_H - V_H - N - D - N - J_H$ unit to one of the 8 C_H gene segments.

All classes of immunoglobulins use the same set of variable region genes and the first B-cells to appear during development carry IgM.

Q. Explain the process of class–switching of immunoglobulin.

Ans. A particular class of immunoglobulin is differentiated from other with respect to its constant regions of the heavy chain. Thus constant region genes, encoding different heavy chains generate antibody class and subclass.

Initially B-cells transcribe a V–D–J gene and a μ-gene for heavy chain. After the removal of introns, mRNA for IgM is produced. During cell maturation, class-switch recombination occurs between the Sμ region and a downstream switch region as each heavy chain gene is preceded by a switch region. At this stage the intervening region is looped out with the deletion of the intervening region and subsequent joining of two switch regions takes place.

Q. What are different antibody variants ?

Ans. There are 3 types of antibody variants. These are referred to as isotypes, idiotypes and allotypes.

1. **Isotypes:** Isotypic variations refer to H and L chain classes and subclasses, *e.g.* γ genes for heavy chains in IgG antibody class are referred to as V_1, V_2, V_3 and V_4, and give rise to IgG isotypes which are designated as IgG_1, IgG_2, IgG_3 and IgG_4, respectively. Genes for isotypic variants are present in all healthy members of a species.

2. **Idiotypes :** Variations within the variable region produce idiotypes. Idiotypes are usually specific for individual B-cell clones. These variations determine binding specificity of the antigen-binding site.

3. **Allotypes :** These variations occur due to constant regions of heavy chains. It refers to genetic variations between individuals within a species.

Q. What is multiple myeloma ?

Ans. Multiple myeloma refers to a neoplastic condition of plasma cells, resulting in large increase of light chain dimers of immunoglobulins. These proteins are referred to as Bence-Jones proteins and are found in the serum and urine of such individuals who are suffering from multiple myeloma.

Q. What are monoclonal antibodies ?

Ans. Benign or malignant proliferation of a single clone, when used as an antigen, results in the production of antibodies which are directed against the binding site for only a particular epitope of an antigen, *i.e.* antibodies of defined specificity produced from a single clone of cells. These are homogenous in nature and are used in the quantitation of several types of proteins, particularly, the nature of an infectious agent.

Q. What are autoimmune diseases ?

Ans. Autoimmune disease refers to the process of loss of immunity against body's own tissues or cells. For example, due to genetic mutation, interaction with drugs, certain types of infections or loss of self-tolerance, immunological appratus of the individual starts forming antibodies against its own tissues. Autoimmune diseases may range from single organ or cell type, such as Grave's disease or IDDM, to a multisystem disease characterised by lesion in several organs, *e.g.* rheumatoid arthritis, SLE, etc.

Q. What is AIDS ?

Ans. AIDS (acquired immunodeficiency syndrome) is caused due to human immunodeficiency virus (HIV). It is characterized by immunosuppression associated with opportunistic infections, secondary neoplasms, and neurological dysfunctions.

Q. What is immunization ?

Ans. Immunization is the introduction of vaccine that leads to protection against virulent form of microorganisms. A wide range of antigen preparations are used as vaccines, *e.g.* rabies vaccine, hepatitis B vaccine, etc.

16

Carcinogenesis and Metabolism of Xenobiotics

Q. Define carcinogenesis.

Ans. Carcinogenesis is the process of development of cancer. An agent which leads to carcinogenesis is referred to as a carcinogen. It may be a chemical, a physical agent such as radiation, or a biological agent, *e.g.*, viruses.

Q. Define mutagenesis.

Ans. Chemical carcinogenesis is referred to as mutagenesis. A direct reacting substance is referred to as a mutagen. When it enters the body, a mutagen may cause DNA damage and mutation.

Q. What are oncogenes ?

Ans. Oncogenes are the DNA sequences carried by viruses, which are capable of causing the transformation of their target cells. Viruses which cause cancer are called oncogenic viruses or tumorogenic viruses. *e.g.,* human papilloma viruse.

Q. What are proto-oncogenes ?

Ans. Proto-oncogenes are the sequences of genes in the host cell that undergo subsequent modifications by retroviruses. A proto-oncogene can be activated by various processes such as point mutation, chromosomal translocation, gene amplification, or insertion of the promotor or enhancer.

Q. What are oncoproteins ?

Ans. Oncoproteins are the products of the oncogenes, *e.g.*, growth factors, growth factor receptors, etc.

Q. What are cancer suppressor genes ?

Ans. Cancer suppressor genes are referred to as antioncogenes. As oncoproteins promote carcinogenesis, the protein products of the antioncogenes apply break to cell proliferation and thus suppress carcinogenesis. Cancer suppressor genes include P_{53}, Rb, etc.

Protein products of the cancer suppress genes have growth

inhibitory effects or include molecules that regulate cell-adhesion.

Q. Name various genetic factors that are responsible for carcinogenesis.

Ans. 1. Activation of oncogenes,

2. Inactivation of tumour suppressor genes,
3. Inactivation of DNA repair genes, and
4. Inactivation of the genes that regulate apoptosis.

Q. What is a xenobiotic ?

Ans. Xenobiotic means a foreign substance which has no biological significance and can enter the body by various routes such as inhalation, ingestion, etc. It is usually water insoluble and may prove toxic, depending upon its level in the body.

Q. What is detoxification ?

Ans. Detoxification, also referred to as detoxication, is the process of the transformation of xenobiotics into their less toxic metabolites.

Q. Describe mechanisms of the metabolism of xenobiotic.

Ans. Xenobiotics are metabolised by phase I and phase II reactions.

Phase I metabolism of xenobiotics include oxidation, reduction or hydrolysis.

Phase II of metabolism involves conjugation of the xenobiotic or its metabolite with glycine, glutamate or some other substance. The conjugated product is excreted since usually it becomes water soluble.

Q. Describe the role of oxidation in the metabolism of xenobiotic ?

Ans. Various compounds such as alcohols, aldehydes, etc., are detoxified by the process of oxidation. Various aromatic hydrocarbons and drugs are also oxidised and their metabolites are excreted in the urine.

Q. Describe the role of cytochrome P_{450} in oxidative metabolism of xenobiotics.

Ans. Cytochrome P_{450} is a class of cytochromes which have absorbance maxima at 450 nm.

1. These are hemoproteins with ferriprotoporphyrin ring.
2. These are located in the smooth endoplasmic reticulum of most tissues.

3. These are monooxygenases which require oxygen and NADPH.

4. These are inducible enzymes and can be synthesized by the cell in response to various chemicals.

5. They oxidise a variety of xenobiotics particularly lipophilic compounds including drugs, carcinogens, pesticides and petroleum products, and results, in the hydroxylation of many substances.

Q. What is cytochrome P_{448} ?

Ans. Cytochrome P_{448}, also referred to as aromatic hydrocarbon hydroxylase, is essentially required for the metabolism of polycyclic hydrocarbons, particularly present in smoke. Smoke have high levels of this enzyme in various tissues, especially lungs.

Q. Briefly describe various conjugation reactions ?

Ans. Phase II reaction of xenobiotic metabolism means either xenobiotic or its metabolite gets conjugated with a randomly available reactant such as glutathione, an amino acid like glycine or glutamine, glucuronate or active sulphate, acetyl or methyl groups.

Conjugation with glucuronic acid : Glucuronic acid (as UDP-glucuronic acid) conjugates with a xenobiotic and forms highly soluble excretory product, *i.e.,* xenobiotic-glucuronate. This reaction is catalysed by the enzyme UDP-glucuronyltransferase. Compounds like benzoic acid, bilirubin, etc., are detoxified after their conjugation as glucuronides.

Conjugation with glutathione : A xenobiotic gets conjugated with glutathione and is excreted as mercapturic acid. (Cystine conjugate of xenobiotic). This reaction is catalyzed by the enzyme glutathione-S-transferase.

Conjugation with glycine : Many aromatic acids such as benzoic acid, cholic acid, etc., are detoxified after their conjugation with glycine.

Benzoic acid + glycine → Benzoylglycine (hippuric acid).

Methylation : Several xenobiotics, particularly those which are derivatives of pyridine, can be detoxified after methylation. Methylation of the xenobiotics requires active methionine by methyltransferase.

17

Organ Function Tests

Q. List important biochemical tests for the assessment of liver functions

Ans. 1. **Serum bilirubin :** normal serum bilirubin concentration is 0.1-1.0 mg/ 100 ml. Most of it is present as unconjugated bilirubin. Increased level of serum bilirubin helps in the diagnosis of jaundice.

2. **Urinary urobilinogen :** Normally, about 1-4 mg of urobilinogen is excreted in the urine daily. Increased excretion of urobilinogen suggest impaired hepatocellular function.

3. **Serum enzymes :** Serum levels of various enzymes are elevated in a liver disease, *e.g.* transaminases, particularly that of glutamate pyruvate transaminase (GPT, also called alanine aminotransferase or ALT) and glutamate oxaloacetate transaminase (GOT, also called aspartate aminotransferase or AST). SGOT and SGPT levels are elevated in viral hepatitis, even before clinical signs and symptoms of the disease appear. Five to 10-fold increase in the levels of these enzymes may also occur in patients with primary or metastatic carcinoma of the liver. On comparing the levels of the two enzymes, SGPT is more liver specific.

Alkaline phosphatase : Serum alkaline phosphatase levels are raised in a hepatobiliary disease. Levels are more markedly raised in extrahepatic obstruction.

Gamma-glutamyltransferase : γ – Glutamyltransferase (γ – GT) is a better index for the diagnosis of alcohol– induced liver disease.

Q. How various liver function tests can be helpful in the differential diagnosis of jaundice ?

Ans. When serum bilirubin exceeds its normal value, it is called hyperbilirubinemia. Most characteristic clinical manifestation of this condition is referred to as jaundice. It is apparent clinically, when serum bilirubin level exceeds 2 mg%.

Jaundice is of 3 types, *i.e.* prehepatic, hepatic and posthepatic.

1. **Prehepatic jaundice :** It is also called as hemolytic jaundice. It is due to the excessive hemolysis of RBCs such as in hemolytic anemia. This, in turn, results in increased unconjugated bilirubin in the serum.

Unconjugated bilirubin may be a result of its increased production such as during hemolysis, reduced uptake by hepatocytes such as in Gilbert's syndrome or decreased biotransformation in the liver such as in neonatal jaundice, Crigler-Najjar syndrome, etc. There is also increased excretion of urobilinogen in the urine of these patients.

2. **Hepatic jaundice :** It is due to a primary liver disease such as viral hepatitis or cirrhosis. Plasma of such individuals show elevated levels of, both unconjugated as well as conjugated bilirubin. In addition, there is also a rise in SGOT and SGPT levels in such individuals.

Hepatic jaundice may also be observed in a benign autosomal recessive condition referred to as Dubin-Johnson syndrome.

3. **Posthepatic jaundice :** It is also called obstructive jaundice since it may be due to obstruction of the bile duct (posthe-patic cholestasis) which may be a result of gallstone, carcinoma of the bile duct or carcinoma of the head of pancreas. Serum of such patients contain increased levels of conjugated bilirubin. Besides, there is also a rise in serum alkaline phosphatase. It is also characterised by the presence of excessive amount of the bilirubin in the urine.

Q. Describe various kidney function tests.

Ans. Commonly used biochemical tests for the assessment of kidney functions include-estimation of blood urea and, serum creatinine, and the detection of abnormal constituents of urine.

Biochemical tests may also be done for the assessment of glo-merular as well as tubular functions, separately.

Commonly used glomerular function tests include, creatinine clearance, urea clearance, rate of renal plasma flow (p-aminohippuric acid clearance), etc.

Assessment of tubular functions can be done by urine concentration dilution or acidification tests.

Q. What is clearance ? Which substance is preferred for its assessment ?

Ans. Clearance is defined as the volume of plasma that is cleared of a particular substance by the two kidneys, per unit time.

It can be calculated, mathematically, by utilising the relationship of a substance in the serum and urine to the rate of flow of urine.

$$C = \frac{U \times V}{P}$$

where C = clearance of the substance (ml/min)

U = concentration of the substance in the urine (mg/dl)

V = flow rate of urine (ml/min)

P = concentration of the substance in plasma, serum or blood (mg/dl)

Various substances which can be used for knowing the clearance include urea, inulin or creatinine.

Inulin clearance : Inulin is a polymer of fructose. It is filtered by the glomeruli but is neither reabsorbed nor secreted by the kidney. Inulin clearance is nearly equal to GFR. However, it is not preferred routinely due to its intravenous administration as it is an exogenous substance and has to be administered.

Urea clearance : Urea is an endogenous substance. It is an end product of protein catabolism, in the liver. However, its clearance depends upon the volume of urine excreted in one minute. Accordingly, urea clearance has to be calculated as standard urea clearance (when the urine volume is below 2 ml/minute) or maximum urea clearance (when urine volume is 2 ml or more/min).

Creatinine clearance : Creatinine is also an endogenous substance, however, its clearance is not affected by the volume of urine. Hence creatinine clearance is preferred as a kidney function test.

Q. List important thyroid function tests.

Ans. Commonly used biochemical tests for the assessment of thyroid function include—

1. **BMR :** BMR is raised in hyperthyroidism and reduced in hypothyroidism

2. **Serum cholesterol :** Increased serum cholesterol concentration is observed in hypothyroidism.

3. **Thyroid hormones :** Thyroxine (T_4), Tri-iodothyronine (T_3) and TSH (thyroid stimulating hormone) concentrations are measured to confirm the diagnosis of hyper or hypothyroidism.

Their normal serum levels are –

Hormone	Serum concentration
T_4	5-12 μg/dl
T_3	70-190 ng/dl
TSH	0.3-0.5 mu/l

18

Free Radicals and Oxygen Toxicity

Q. What are free radicals or reactive oxygen species ?

Ans. The elemental oxygen (O_2), as we all know, is essential for life. About 90% of it is used via oxidative phosphorylation while the other 10% is used in enzymatic hydroxylation and oxygenation reactions. A small fraction of the oxygen is also converted to inherently toxic forms, such as superoxide and hydrogen peroxide. These are the highly reactive forms of oxygen and are referred to as free radicals or reactive oxygen species (ROS). They are a source of chronic damage to biomolecules, within the tissues.

In biological redox reactions involving O_2, oxygen is activated by the redox active metal ions, such as iron and copper, which are present in the metallo-enzymes and other proteins, e.g. hemoglobin, myoglobin, etc. These metal ions provide one electron at a time to oxygen and activate it to strong oxidizing forms. A number of reactive oxygen species, also called free radicals, are formed from the partial reduction of molecular oxygen. The first product is the anion radical called superoxide ($O_2^{\cdot}$). Reduction of superoxide yields hydroperoxide, in the form of H_2O_2. Reduction of H_2O_2 causes a hemolytic cleavage reaction that releases hydroxyl radical ($OH^{\cdot}$). The end product of complete reduction of oxygen is water.

ROS are formed by three mechanisms in vivo, i.e. by reaction of oxygen with decompartmentalised metal ions, as a side reaction of electron transport, or by normal enzymatic reaction of electron transport, e.g. monoamine oxidase (in dopamine metabolism), fatty acid oxidases (in the peroxisomes), etc.

Q. What do you mean by oxidative stress ?

Ans. Free radicals or reactive oxygen species, as we know, are continuously formed as the by-products of aerobic metabolism through reactions with drugs and environmental toxins and are utilized by the antioxidants that are present in the cell. When either there is increased production of the free radicals or the levels of antioxidants are diminished, this in turn results in a situation called **oxidative stress**.

These reactive oxygen species can cause serious chemical damage to the cellular macromolecules, such as DNA, proteins and unsaturated lipids, and can lead to cell death. They have been implicated in a number of pathological processes, including reperfusion injury, cancer, inflammatory disease and aging.

Q. Explain the nature of damage that can be caused by free radicals.

Ans. Amongst the various free radicals, hydroxyl radical ($OH^{\bullet}$) is the most reactive and damaging. Its half life is only few nanoseconds. It reacts with biomolecules primarily by hydrogen abstraction and addition reactions. One of the most sensitive sites of free radical damage is the cell membrane, which is rich in readily oxidized polyunsaturated fatty acids.

The hydroxyl radical abstracts a hydrogen atom from a polyunsaturated fatty acid, setting off a chain of lipid peroxidation reactions. Lipid peroxides formed in this reaction degrade to form characteristic products, such as malondialdehyde (MDA). These compounds react with proteins to form adducts and cross links, known as advanced lipoxidation end-products (ALE).

ROS also react with carbohydrates to form dicarbonyl compounds that react with protein to form cross links and adducts, known as glycoxidation products or advanced glycation end-products (AGE).

ROS also cause some residual damage to the genome, in the form of mutations in DNA.

As a result of oxidative stress, MDA adducts, ALE and AGE, increase in several chronic conditions, such as atherosclerosis, Alzeheimer's disease and diabetes.

Q. What are the antioxidant defenses present in cells ?

Ans. There are several levels of protection against oxidative damage that a cell has. These include antioxidant enzymes and vitamins.

Enzymatic Defenses

There are a group of enzymes that act to detoxify the precursors of hydroxyl radicals. These include catalase, superoxide dismutase (SOD) and glutathione peroxidase (GPx_x).

Catalase detoxifies H_2O_2 and is found in large amount in peroxisomes:

$$2H_2O_2 \xrightarrow{\text{Catalase}} O_2 + 2H_2O$$

SOD detoxifies superoxide to the less toxic H_2O_2:

$$O_2^{\cdot} \xrightarrow{\text{SOD}} O_2 + H_2O$$

There are two classes of SOD. One is Mn.SOD isoenzyme, which is found in the mitochondria while the other is Cu-Zn.SOD, which is widely distributed in all the compartments of the cell.

Glutathione peroxidase (GPx) reduces both, H_2O_2 and lipid peroxides to water and a lipid alcohol, respectively. GPx requires reduced glutathione (GSH), which acts as a co-substrate.

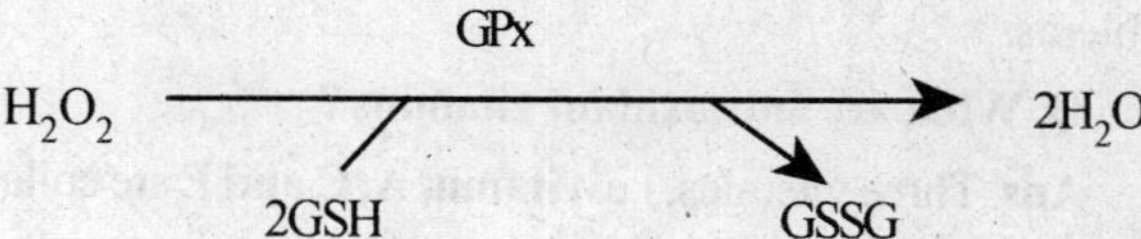

GSH is a tripeptide (γ-glutamylcysteinylglycine) containing glutamate, cysteine and glycine. It is found in all the cells and chemically detoxifies H_2O_2. This reaction is catalyzed by GPx, which is a selenium-containing enzyme. In this reaction, GSH is oxidized to GSSG (oxidized glutathione), which has no protective effect. The cell regenerates GSH from GSSG by an NADPH-dependent enzyme called glutathione reductase. NADPH is produced by the pentose phosphate pathway.

ANTIOXIDANT DEFENSES IN ERYTHROCYTES

Free radicals are also continuously formed in the erythrocytes and initiate lipid peroxidation reaction, which can lead to loss of membrane integrity, causing cell lysis. The RBCs are well protected against the oxidative stress, as the cell has a high concentration of GSH. Moreover, the pentose phosphate pathway, which supplies NADPH, is the major pathway that is operative in the erythrocytes.

ANTIOXIDANT VITAMINS

Three vitamins, i.e. **vitamin A, C and E**, provide the other line of defense against oxidative damage and are called antioxidant vitamins.

Vitamin C (ascorbate) in the aqueous phase while **vitamin E** (α- and γ-tocopherols) in the lipid phase, act as chain-breaking antioxidants. Vitamin C reduces superoxide and lipid peroxyl radicals. It also has a role in reduction and recycling of vitamin E. In response to severe oxidative stress, vitamin E is maintained at constant concentration in the lipid phase until all the vitamin C has been consumed. Thus, these vitamins do not propagate radical damage and are recycled.

Vitamin A is a lipophilic antioxidant. It is a potent singlet oxygen scavenger and protects the cell against damage from sunlight, particularly the retina and the skin.

Consumption of foods rich in ascorbate, vitamin E and β-carotene, has been correlated with a reduced risk for certain types of cancer and decreased frequency of other chronic health problems.

Q. What are anti-oxidant vitamins ?

Ans. Three vitamins, i.e. **vitamin A, C and E** are collectively called as antioxidant vitamins.

Vitamin C (ascorbate) in the aqueous phase while **vitamin E** (α-and γ-tocopherols) in the lipid phase, act as chain-breaking antioxidants. Vitamin C reduces superoxide and lipid peroxyl radicals. It also has a role in reduction and recycling of vitamin E. In response to severe oxidative stress, vitamin E is maintained at constant concentration in the lipid phase until all the vitamin C has been consumed. Thus, these vitamins do not propagate radical damage and are recycled.

Vitamin A is a lipophilic antioxidant. It is a potent singlet oxygen scavenger and protects the cell against damage from sunlight, particularly the retina and the skin.

Consumption of foods rich in ascorbate, vitamin E and β-carotene, has been correlated with a reduced risk for certain types of cancer and decreased frequency of other chronic health problems.

19

Extracellular Matrix

Q. What is extra-cellular matrix ?

Ans. Extra-cellular matrix includes various types of connective tissues, including tendons, cartilage and the corneal stroma. Major components of the extra-cellular matrix include **collagens, elastin, proteoglycans** and a variety of other proteins. Each of these proteins of the extra-cellular matrices has a modular construction composed of domains that contain binding sites for one another and for receptors on the cell surface. As a result, these various extra-cellular materials interact to form an interconnected network that is bound to the cell surface.

Q. What are collagens? How they are synthesized? Describe some abnormalities of collagen biosynthesis.

Ans. Collagens comprise a family of **fibrous glycoproteins** that are present only in the extra-cellular matrices. Collagen is the most abundant protein in the human body. It has a long rigid structure in which **three polypeptides**, called **α-chains**, are wound around one another, in a **rope-like triple helix**. Collagens are found throughout the body. Their types and organization are determined by the structural role collagen plays in a particular organ.

In some tissues, collagen may be dispersed as a gel that gives support to the structure, as in the extra-cellular matrix or the vitreous humor of the eye.

In other tissues, collagen may be bundled in tight, parallel fibers that provide great strength, as in tendons.

In the cornea of the eye, collagen is stacked so as to transmit light with a minimum of scattering.

Collagen of bone occurs as fibers arranged at an angle to each other so as to resist mechanical sheer from any direction.

TYPES OF COLLAGEN

It is a super-family of proteins, which include 27 distinct types. Although there are many differences among the members of the collagen family, all share at least two important structural features:

1. All collagen molecules are **trimers** consisting of three polypeptide chains, called α-chains.

2. The **three polypeptide chains** of a collagen molecule are **wound around each other** to form a unique, **rod-like triple helix**, e.g. collagen type I is $\alpha 1_2 \alpha 2$, i.e. it has two $\alpha 1$ and one $\alpha 2$ chains. Type II collagen is $\alpha 1_3$, i.e. it contains 3 $\alpha 1$ chains, etc.

Based on their location and functions in the body, collagens are organized into 3 groups:

1. **Fibril-forming Collagens** – These are linear polymers of fibrils with characteristic bonding patterns and have rope-like triple-stranded helix, e.g. type I, type II and type III. Type I collagen fibers are found in supporting elements of high tensile strength, such as in skin, bone, tendon, blood vessels and cornea. Type II fibers are found in cartilaginous structures, such as cartilage, inter-vertebral disk and vitreous body. Type III collagens occur in more distensible tissues, such as blood vessels and fetal skin.

2. **Network forming Collagens** – They form a three-dimensional mesh, e.g. type IV and VII collagens. Type IV collagens are thin, sheet or mesh-like structures that function as semi-permeable basement membranes found in the kidney and the lung. Type VII collagens are found beneath stratified squamous epithelia.

3. **Fibril-associated Collagens** – These collagens link collagen fibrils to one another as well as to other components in the extra-cellular matrix, e.g. type IX and type XII collagens. Type IX collagens are found in cartilage whereas type XII collagens occur in tendon and ligaments.

STRUCTURE AND COMPOSITION

Collagen is a fibrous protein, which has an elongated triple-helical structure. It is **rich in proline and glycine**, both of which are important in the formation of the triple-stranded helix. Proline facilitates the formation of the helical conformation of each α-chain because of its ring structure that causes kinks in the peptide chain. **Glycine** is found in **every third position** of the polypeptide chain, forming part of a repeating sequence **Gly-X-Y-**. It may be due to the reason that glycine is the smallest amino acid and fits into the restricted space where three chains of the helix come together. **X is frequently proline while Y may be proline or lysine**, which **get hydroxylated** and occur as hydroxyproline or hydroxylysine.

BIOSYNTHESIS

1. The polypeptide precursors of the collagen molecules are **formed in fibroblasts** (or osteoblasts of bone and chondroblasts of cartilage) and **secreted into the extra-cellular matrix**. The newly synthesized polypeptide precursor of α-chains contains a special amino acid sequence at the N-terminal end. This signal sequence facilitates the binding of ribosomes to the rough endoplasmic reticulum and directs the passage of the polypeptide chain into the cisternae of the RER. The signal sequence is rapidly cleaved in the endoplasmic reticulum to yield a precursor called **pro-α-chain**.

2. The pro-α-chain is processed by a number of enzymes within the lumen of the RER while the polypeptides are still being synthesized.

Proline and lysine residues found in the Y-position **are hydroxylated** by prolyl hydroxylase and lysyl hydroxylase, respectively. This is a post translational modification and the hydroxylation reaction requirse molecular oxygen and vitamin C. Hydroxyproline is important in stabilizing the triple-helical structure of collagen because it maximizes inter-chain hydrogen bond formation. In case of **vitamin C deficiency**, collagen **fibers cannot be cross-linked**, greatly **decreasing the tensile strength** of the assembled fiber, such as in scurvy.

3. **Some hydroxylysine residues are** enzymatically **glycosylated** with glucose or glycosyl-galactose.

4. After hydroxylation and glycosylation, pro-α-chains form **pro-collagen**. Pro-collagen molecules are translocated to the Golgi apparatus where they are **packed in secretory vesicles**. The vesicle fuses with the cell membrane, causing the release of pro-collagen molecules into the extra-cellular space.

5. After their release, the pro-collagen molecules are cleaved by N- and C-pro-collagen peptidases, which remove the terminal propeptides, releasing **triple-helical molecules**.

6. Individual collagen molecules spontaneously associate to **form fibrils**.

7. Lysyl oxidase, oxidatively deaminates some of the lysyl and hydroxy-lysyl residues, resulting in the **formation of allysine and hydroxy-allysine**. These reactive aldehydes condense with lysyl or hydroxy-lysyl residues in neighboring collagen molecules to **form covalent cross-links.**

DEGRADATION

Normal collagens are highly stable molecules. However, connective tissue is constantly being remodeled, often in response to growth or injury of the tissue. Collagen fibrils are broken down with the help of specific metalo-proteinases, called **collagenases**. These enzymes cleave the molecule into smaller fragments, which are further degraded to the constituent amino acids by other matrix proteinases.

ABNORMALITIES IN COLLAGEN SYNTHESIS

A large number of mutations have been identified in genes coding for collagen protein or processing enzymes. The following diseases can occur as a result of defective collage synthesis:

1. Mutations in the gene for either the pro 1- or pro 2- α chains of type I collagen can result in structurally abnormal pro α -chains that can prevent folding of the protein into a triple-helical conformation. This is called **osteogenesis imperfecta** (OI), also known as brittle bone syndrome. Extremely fragile bones, thin skin and weak tendons characterize it. It is a heterogeneous group of inherited disorders distinguished by bones that easily bend and fracture.

Type I osteogenesis imperfecta, called osteogenesis imperfecta tarda, is present in early infancy with fractures secondary to minor trauma.

Type II osteogenesis imperfecta, called osteogenesis imperfecta congenita is more severe and patients die in utero or in the neonatal period of pulmonary hypoplasia.

2. A deficiency of any collagen processing enzymes, e.g. lysyl hydroxylase or procollagen peptidase, results in **Ehlers-Danlos syndrome**. This disorder is a heterogeneous group of generalized connective tissue disorders that result from inheritable defects in the metabolism of fibrillar collagen molecules. These patients show defects in collagen type I and type III fibrils. They have hyperflexible joints and highly extensible skin.

3. Mutation in type IV collagen genes occur in-patients with **Alport syndrome**, an inherited kidney disease in which the glomerulus basement membrane is disrupted.

Q. What is elastin ? Describe its structure. Write a disorder of elastin biosynthesis.

Ans. Elastin is a connective tissue protein with rubber-like properties. It is the predominant protein of the elastic fibers in the extra-cellular matrices of blood vessels, lungs, ligaments and spine.

STRUCTURE AND SYNTHESIS

Elastin is a polymer synthesized from a precursor called **tropo-elastin**. Tropo-elastin is a linear polypeptide synthesized on the rough endoplasmic reticulum and secreted by the cell into the extra-cellular space. Tropo-elastin does not undergo post-translational modification. During the assembly process into extra-cellular space, some of the lysyl side chains of the tropo-elastin polypeptides are oxidatively deaminated by lysyl oxidase, generating allysine in specific sequences. The reactive aldehyde of allysine condenses with other allysine or with unmodified lysines, from the same or on different tropo-elastin polypeptide chains, to form **desmosine cross-link heterocyclic structures**. As a result, elastin can stretch in any direction giving elasticity to the connective tissue.

Elastin is **rich in glycine, proline, lysine and valine**. One in seven of its amino acids are valine. Its primary structure consists of alternating hydrophilic and hydrophobic, lysine and valine-rich domains. Lysines are involved in intermolecular cross-linking while the weak interactions between valine residues in the hydrophobic domains impart elasticity to the molecule.

Elastin is an insoluble, polymeric, amorphous core covered with a sheath of microfibrils, the predominant constituent of which is the **glycoprotein fibriltin**.

DISORDERS OF ELASTIN METABOLISM

Mutations in the fibriltin gene results in a genetic disease of connective tissues, called **Marfan syndrome**. People with the disease have typically tall stature, long arms and legs, and long spidery fingers. The disease in a mild form causes loose joints, deformed spine, floppy mitral valves, and eye problems, such as lens dislocation. In severely affected individuals, the aorta wall is prone to rupture because of defects in elastic fiber formation.

Elastin is degraded by the enzyme **elastase**. This enzyme is a powerful protease that is released into the extra-cellular space and degrades elastin of alveolar walls and other structural proteins in a variety of tissues. **α_1-Antitrypsin** (α_1-AT), produced primarily by liver and other tissues, such as monocytes and alveolar macrophages, **inhibits** a number of proteolytic enzymes including **elastase**. This in turn prevents elastin degradation in the alveolar walls. A **deficiency of α_1-AT can cause emphysema,** which results from the destruction of the connective tissue of alveolar walls.

Q. What are glycosaminoglycans? Describe structure, distribution and functions of various types of glycosaminoglycans.

Ans. Glycosaminoglycans, originally designated as **mucopolysaccharides,** are large complexes of negatively charged hetero-polysaccharide chains associated with a small amount of protein, forming proteoglycans, which are the gel forming components of the extra-cellular matrices.

Glycosaminoglycans are **unbranched polysaccharides** and consist of **alternating uronic acid and hexosamine residues**. Solutions of glycosaminoglycans have slimy, mucus like consistency that results from their high viscosity and elasticity

FUNCTIONS OF GLYCOSAMINOGLYCANS

1. One of their major roles is to provide **structural support to** tissues, especially **cartilage and connective tissue**.

2. Extra-cellular matrix enables the tissue to **withstand torsion and shock**.

3. Cartilage provides a balance between **integrity and flexibility**.

4. They also perform **specific functions in certain organs**, such as liver, kidney, cornea, etc. Heparan sulfate mediates the attachment of hepatocytes to their substratum in the cell culture.

Changes in both the collagen and proteoglycan content of the renal basement membrane are associated with diabetic renal disease.

Corneal clouding in macular corneal dystrophy is associated with under-sulfation of keratin sulfate I.

Heparin and heparan sulfate have physiological roles in binding proteins or other macromolecules, such as proteinases in mast cells, LDL and lipoprotein lipase to the vascular wall.

STRUCTURE AND CLASSIFICATION OF GLYCOSAMINOGLYCANS

Each glycosaminoglycan chain is composed of a **repeating disaccharide unit**, which comprises of **an acidic sugar and an amino sugar**, i.e. it has the structure as –

[Acidic sugar – amino sugar]$_n$

The acidic sugar is either D-glucuronic acid or its carbon 5 epimor, called L-iduronic acid.

The amino sugar is either D-glucosamine or D-galactosamine in which the amino group is usually acetylated. It may also be sulfated on carbon 4 or 6, or on non-acetylated nitrogen.

Glycosaminoglycans are divided into **six major classes**, according to monomeric composition, type of glycosidic linkages and degree and location of sulfate units. The general structures of glycosaminoglycans and their distribution are shown in Table.

Table: Structure, Distribution and Functions of Glycosaminoglycans

Glycosaminoglycan	Repeating disaccharide unit	Tissue Location	Major functions
Hyaluronic acid	[GlcUA-GlcNAc]	Joint and ocular fluids	Serves as excellent biological shock absorber and lubricants
Chondroitin sulfates	[GlcUA-GlcNAc]	Cartilage, tendons, bone	Bind collagen and give it strength
Dermatan sulfate	[IdUA-GalNAc]	Skin, valves, blood valves	
Heparan	[IdUA-GlcNAc]	Cell surfaces	Initiation of

sulfate			signaling processes
Heparin	[IdUA-GlcNAc]	Mast cells, liver	Serves as an anticoagulant.
Keratan sulfates	[Gal-GlcNAc]	Cartilage, cornea	

GlcUA – Glucuronic acid

GlcNAc – N-Acetylglucosamine

GalNAc – N-Acetylgalactosamine

IdUA – Iduronic acid

Gal – Galactose

Hyaluronic Acid

Hyaluonic acid is composed of **repeating** units of **D-glucuronic acid and N-acetyl-D-glucosamine**. This polysaccharide chain is the longest of the glycosaminoglycans (GAG) and differ from other types, as it is the only GAG, which is non-sulfated. It is not covalently attached to protein and is not limited to animal tissue but also found in bacteria. It is an important GAG component of **connective tissue, synovial fluid and the vitreous humor** of the eye. The visco-elastic behavior of hyaluronate solution makes them excellent biological shock absorbers and lubricants.

Chondroitin Sulfates

Chondroitin sulfates contain **glucuronic acid and N-acetylgalatosamine with sulfate** on the hydroxyl group of either carbon 4 (chondroitin-4-sulfate) or carbon 6 (chondroitin-6-sulfate). They are the major components of cartilage where they bind collagen and hold fibers in a tight, strong network. They are also found in **aorta, tendons and ligaments.**

Dermatan Sulfate

Dermatan sulfate is derived from chondroitin, by enzymatic epimerization of carbon 5 of glucuronate residues, to form iduronate residues. Thus, dermatan sulfate contains **L-iduronic acid** (with variable amounts of glucuronic acid) and **N-acetylgalactosamine**. It is found in **skin, blood vessels, and tendon and heart valves.**

Heparan Sulfate

Heparan sulfates are not constructed of identical disaccharides but exhibit structural diversity. They consist of **glucuronic or iduronic acid and glucosamine**. Some glucosamines are acetylated and may have N- and O-sulfate groups. It is a ubiquitous cell-surface component as well as extra-cellular substance in **blood vessel walls and brain**. Heparan sulfates, which function directly at the cell surface, interact with a variety of proteins, including important growth factors and their receptors (such as fibroblast growth factors and their receptors). The formation of ternary complexes of heparan sulfate with fibroblast growth factor and fibroblast growth factor receptor initiates signaling processes.

Heparin

Heparin consists of **glucuronic or iduronic acid and glucosamine**. Most of its glucosamine units are N-sulfated. Sulfate is also bound to the hydroxyl group on carbon 2 of the uronic acid residues and the carbon 3 or 6 of glucosamine. It is a highly charged polymer, found in the **intracellular granules of the mast cells** that occur in arterial walls, especially **in liver, lungs and skin**. Heparin is widely used, clinically, to **inhibit blood clotting**, e.g. in post-surgical patients.

Keratan Sulfates

Keratin sulfate is the most heterogeneous group of glycosaminoglycans as its sulfate content is variable and it contains small amounts of **fucose, mannose, N-acetylglucosamine and sialic acid**. The repeating disaccharide, generally, is N-acetyl-glucosamine and galactose. Sulfate content is variable and may be present on the carbon 6 hydroxyl groups. Keratan sulfates are linked to protein either by an N-linked (keratan sulfate I) or O-linked (keratan sulfate II) oligosaccharides. Keratan sulfate I is found in **cornea** while keratan sulfate II occurs in **loose connective tissue** proteoglycan aggregates with chondriotin sulfate.

SYNTHESIS OF GLYCOSAMINOGLYCANS

They are synthesized by a series of glycosyl transferases, epimerases and sulfotransferases, beginning with the synthesis of the core oligosaccharides while the core protein is still in the rough endoplasmic reticulum. Synthesis of the repeating oligosaccharide and other modifications take place in the Golgi apparatus.

The polysaccharide chains are elongated by **sequential addition of alternating acidic and amino sugars**, donated by their UDP-derivatives. UDP-glucuronic acid is produced by oxidation of UDP-glucose. After D-glucuronic acid has been incorporated into the carbohydrate chain, some of the D-glucuronic acid residues are converted to L-iduronic acid residues. Epimerization of the D- to the L-sugar is caused by the enzyme uronosyl-5-epimerase.

Sulfation of the carbohydrate chain occurs after the monosaccharide (to be sulfated) has been incorporated into the growing oligosaccharide chain. 3-Phosphoadenosyl-5′-phosphosulfate (PAPS) is the source of the sulfate, which is transferred at specific sites to cause sulfation of the carbohydrate, by sulfo-transferases.

DEGRADATION OF GLYCOSAMINOGLYCANS

The glycosaminoglycan chains are degraded by the sequential action of a number of different **lysosomal acid hydrolases**. Their stepwise degradation involves exo-glycosidases and sulfatases, beginning from the external end of the glycan chain. This may involve the removal of sulfate by a sulfatase, then removal of the terminal sugar by a specific glycosidase, and so on.

DEFECTS OF GLYCOSAMINOGLYCAN DEGRADATION

If one of the enzymes involved in the stepwise pathway is missing, the entire degradation process is halted at that point and the un-degraded molecules accumulate in the lysosome. The lysosomal storage diseases resulting from the accumulation of these un-degraded molecules are known as mucopolysaccharidoses.

All the deficiencies are autosomal and recessively inherited except Hunter syndrome, which is X-linked. They are characterized by the accumulation of glycosaminoglycans in various tissues, causing varied symptoms, such as skeletal and extra-cellular matrix deformities and mental retardation. Children who are homozygous for one of these diseases are apparently normal at birth and then gradually deteriorate. In most severe cases, death occurs in childhood.

In general, these diseases can be diagnosed by the identification of the specific glycosaminoglycan chains in the urine, followed by the assay of the specific hydrolases in leukocytes or fibroblasts.

Q. Write a note on lysosomal storage diseases.

Ans. Lysosomes are responsible for intracellular digestion of the extra-cellular and the intracellular substances. They contain a number of hydrolyzing enzymes (hydrolases) for the hydrolysis of proteins, lipids, carbohydrates, and nucleic acids. These hydrolases are most active at acidic pH and split complex molecules into simple low molecular weight compounds that can be reutilized.

In a number of genetic disorders, classified as lysosomal storage diseases, individual lysosomal enzymes are missing, leading to accumulation of the substrate of the missing enzyme. The lysosomes become enlarged with undigested material, thereby interfering with the normal cell processes. Several lysosomal storage diseases are known to exist, e.g. **sphingolipidoses, mucopolysaccharidoses**, etc.

MUCOPOLYSACCHARIDOSES

Mucopolysaccharidoses are caused by a deficiency of one of the lysosomal hydrolases, which are normally involved in the degradation of heparan sulfate and/or dermatan sulfate, e.g. **Hurler syndrome and Hunter syndrome**. These are the hereditary disorders and are characterized by the accumulation of glycosaminoglycans in various tissues, causing varied symptoms, such as skeletal and extra-cellular matrix deformities and mental retardation. This also results in the excretion of oligosaccharides in the urine because of the incomplete lysosomal degradation of glycosaminoglycans. Some of the disoreders are shown in table.

The Mucopolysaccharidoses

Syndrome	Deficient enzyme	Product accumulated in lysosomes and excreted in urine
Hunter syndrome	Iduronate sulfatase	Heparan sulfate and dermatan sulfate
Hurler syndrome	α-L-Iduronidase	Heparan sulfate and dermatan sulfate

Morquio syndromes			
	A	Galactose-6-sulfatase	Keratan sulfate
	B	β-Galactosidase	
Sanfilippo syndrome			
	A	Heparan sulfamidase	Heparan sulfate
	B	N-acetylglycosaminidase	
	C	N-Acetylglucosamine -6-sulfatase	
Sly Syndrome		β-Glucuronidase	Dermatan sulfate and heparan sulfate

Q. What are sphingolipidoses ?

This is a group of lipid storage diseases caused by the deficiency of the specific lysosomal hydrolytic enzyme that causes **defects in the degradation of sphingolipids**. Usually, only a single sphingolipid, which is the substrate for the deficient enzyme, accumulates in the involved organs in each disease, such as Niemann-Pick disease, Gaucher's disease and Tay-Sach disease. These are autosomal recessive diseases that can be diagnosed by measuring enzyme activity in cultured fibroblasts or peripheral leucocytes.

Niemann-Pick disease

It results from a **deficiency of** the enzyme **acid sphingomyelinase**, which leads to accumulation of sphingomyelin in lysosomes. It is characterized by hepatosplenomegaly.

Gaucher's disease

It is a lysosomal storage disease where the enzyme **gluco-cerebrosidase is missing** in the afflicted individuals. It is characterized by hepatomegaly, osteoporosis of long bones and neurodegenerative involvement.

Tay-Sach's disease

It is due to **lack of** the enzyme **hexosaminidase**. It is the most common form of GM_2 gangliosidosis. Ganglion cells of the cerebral cortex become swollen due to the deposition of GM_2 gangliosides. Neurodegeration, blindness, cherry-red macula, muscular weakness, and seizures characterize it.

Q. Write a note on osteogenesis imperfecta.

Ans. Mutations in the gene for either the pro 1- or pro 2- α chains of type I collagen can result in structurally abnormal pro α-chains that can prevent folding of the protein into a triple-helical conformation. This is called **osteogenesis imperfecta** (OI), also known as brittle bone syndrome. Extremely fragile bones, thin skin and weak tendons characterize it. It is a heterogeneous group of inherited disorders distinguished by bones that easily bend and fracture.

Type I osteogenesis imperfecta, called osteogenesis imperfecta tarda, is present in early infancy with fractures secondary to minor trauma.

Type II osteogenesis imperfecta, called osteogenesis imperfecta congenita is more severe and patients die in utero or in the neonatal period of pulmonary hypoplasia.

Q. Write a note on Ehlers-Danlos syndrome.

Ans. A deficiency of any collagen processing enzymes, e.g. lysyl hydroxylase or pro-collagen peptidase, results in **Ehlers-Danlos syndrome**. This disorder is a heterogeneous group of generalized connective tissue disorders that result from inheritable defects in the metabolism of fibrillar collagen molecules. These patients show defects in collagen type I and type III fibrils. They have hyper-flexible joints and highly extensible skin.

Q. What is Marfan syndrome ?

Ans. Mutations in the fibriltin gene results in a genetic disease of connective tissues, called **Marfan syndrome**. People with the disease have typically tall stature, long arms and legs, and long spidery fingers. The disease in a mild form causes loose joints, deformed spine, floppy mitral valves, and eye problems, such as lens dislocation. In severely affected individuals, the aorta wall is prone to rupture because of defects in elastic fiber formation.

Q. What is Tay-Sach's disease ?

Ans. It is due to **lack of** the enzyme **hexosaminidase**. It is the most common form of GM_2 gangliosidosis. Ganglion cells of the cerebral cortex become swollen due to the deposition of GM_2 gangliosides. Neuro-degeneration, blindness, cherry-red macula, muscular weakness, and seizures characterize it.

20

Muscle Contraction

Q. Describe the structure and composition of skeletal muscle.

Ans. Muscles form a large proportion of the **active tissues** of the body. In a normal adult, though muscle mass is nearly **40% of the body weight,** metabolically it is a highly active tissue and performs more than 50% of the total metabolic activities.

Structure of Skeletal Muscle

The **voluntary muscles** which include the **skeletal muscle** have a **stariated (stripped) appearance.** Such muscles consist of **long, thin, multinucleated cells** known as **muscle fibres that** run along the length of the muscle.

The muscle fibres are nearly 0.01 to 0.1 nm in diameter and 1 to 40 nm in length. A large number of these fibres are **bound together by connective tissue** and are attached via tendons to the bone **or other structures whose movement** they control. Muscle fibres are packed in an **elastic sheet** known as **sarcolemma** which also has several **nuclei** and **mitochondria.**

The sarcolemma is an **electrically excitable membrane** and is important in the **transport of motor nerve impulses. It also contains** a large number of **transport ducts** called sarcotubules. Sarcolemma also contains **sarcoplasmic reticulum,** similar in structure and functions to the endoplasmic reticulum found in other cells.

Bundles of muscle fibres are surrounded by the **sarcoplasm** which may be called as **muscle plasma.** Sarcoplasm is **semifluid** in nature and contains **myoglobin,** similar to haemoglobin present in red blood cells.

Myoglobin transports oxygen from blood capillaries to the site of oxidation.

MYOFIBRILS

Muscle fibres contain parallel bundles of myofibrils (Greek: myos means muscle). Major features of the myofibril are that, it has a light band called as the isotropic band(I band) which contains thin filaments and a dark band, designated as anisotropic band (A band) which has a central H Zone containing thick filaments and an outer segment with the over-lapping thick and thin filaments. I band is bisected by a dark zone (present in the centre of I band), called as the Z line.

The functional unit of a myofibril is known as sarcomere (Greek: sarkos means flesh). It is a region between two successive Z lines (present at the center of each I band), i.e. the region comprising of one-half of I band, the A band and one half of the next I band.

A band is formed of a large number of thick filaments whereas I band is composed of a comparable number of thin filaments which are interspaced between the thick filaments. Two sets of the filaments, in turn, are linked by cross-bridges in the area where they overlap.

Thin filament slide over the thick filaments, as a result muscle contraction leads to a decrease in length of the sarcomere, caused by a reduction in distance between the I-band and the H-zone. The lengths of the thick filaments and the thin filaments however, remain constant. A contracted muscle thus is reduced to nearly two-third of its fully extended length.

Composition of the Skeletal Muscle

Adult muscle is 70-80% water.

Besides, muscle also contains nearly 20% proteins and 3% lipids (with variable amounts of triacylglycerol, cholesterol and phospholipids).

Muscle also contains glycogen. Glycogen content of the muscle varies from 0.5 to 1.0%. This in turn serves as a source of ATP via anaerobic glycolysis.

Composition of the Skeletal Muscle

Ingredients	Percentage of total muscle mass
Water	70-80%
Proteins	20%
Myosin	
Actin	
Actin cross-linking proteins	
Tropomyosin	
Troponin	
Myoglobin	
Lipids	3%
Triacylglycerol	
Cholesterol	
Phospholipids	
Glycogen	0.5-1.0%
Other substances	
Inositol (Myoinositol)	
Creatinephosphate	
Oligopeptides (carnosine, anserine and glutathione)	
Free amino acids	
Enzymes	
Carnitine	
Inorganic ions	
Cations (Na^+, K^+, Mg^{2+})	
Anions (Phosphates, sulphates and bicarbonate)	

Muscle tissue also contains several other substances, such as inositol, carnitine, free amino acids, various enzymes and several cations (K^+, Na^+ and Mg^{2+}) as well as anions (phosphates, sulphates and bicarbonates). Though several isomers of inositol are found in the muscle, predominant of these is myoinositol.

In addition various small peptides such as carnosine, anserine, β-alanyl-3-methylhistidine and glutathione are also found in the muscle.

Q. Describe various functions of proteins present in the skeletal muscle.

Ans. More than half of the total proteins, present in the skeletal muscle, are structural proteins which include myosin, actin, actin cross-linking proteins tropo-myosin and troponin. Myosin and actin together, account for more than two-third of the total muscle proteins. Muscle also contains myoglobin, several enzymes, collagen and some proteins of the connective tissue.

Myosin

Thick filaments are composed, almost entirely of myosin, which are placed 'end-to-end' with the light chains, striking out in the form of double heads.

Myosin consists of six polypeptide chains which include two identical heavy chains and two pairs of two different types of the light chains.

Two types of the light chains are described as regulatory light chains and essential light chains.

Each myosin head associates with an essential light chain and a regulatory light chain.

Regulatory light chains play a role in the control of muscle contraction by Ca^{2+} ions whereas the essential light chains are required for ATPase activity of the myosin head. Thus, both types of light chains are involved in the regulation of muscle contraction.

The N-terminal half of each of the heavy chain forms an elongated globular head to which one subunit, each of the two light chains binds. The C-terminal half of the heavy chain forms a long fibrous α-helical tail, two of which associate to form a left-handed coil. Taken together, a molecule of myosin, thus, consists of a long rod like segment with two globular heads.

Actin

Actin is a globulin and is a major constituent of thin filaments of the sarcomere. Actin exists in two forms, i.e. as G actin and as filament actin (F actin).

G actin is a low molecular weight mono-meric protein (molecular weight nearly 42 KD) with a globular configuration. These monomers readily form large, elongated actin filament polymers called as F actin.

For assembly of the G actin mono-meric units into a polymeric F actin, Mg^{2+} ions are essential.

Besides mono-meric G actin units are oriented in such a way that they acquire a pointed-region (pointed-end) and a barbed-region (barbed-end). Assembly of the mono-meric units occurs at the barbed-end. A group of proteins called parafilins binds to G actin, stabilizes the monomers and prevents their polymerisation.

F actin thus is a polymerized form of G actin. It contains nearly 1000 monomeric units which are organized as a string of beads. F actin filaments are made up of two strings of the G actin monomers.

Actin is ubiquitous and is usually a most abundant cytoplasmic protein in eukaryotic cells, typically accounting for 5 to 10% of the total protein. Though actin is most prominent in the muscle, it also exists in other tissues where it is responsible for various motile processes, such as cell locomotion, cytoplasmic streaming and cytokinesis (separation of the daughter cells, during cell division). The non-muscle actin forms fibres which are designated as microfilaments. Along with the other protein fibres, these microfilaments form elements of the cytoskeleton. Assembly of the actin monomers, into the fibres, requires ATP.

Actin Cross-Linking Proteins

There are some proteins that cross-link and stabilize actin chains. These include filamin, α-actinin and β-actinin.

Filamin : It is a high molecular weight, flexible protein that usually exists as a dimer. It binds to actin at nearly 40 nm intervals.

α-Actinin: It is mainly located in Z line of the muscle fibres and has a role in spacing of the actin filaments.

β-Actinin : It acts as a capping protein and thus is also called as acting capping protein. It binds to the active site of the G actin monomer, so that the monomer is unable to combine with another monomer. It thus stabilises G monomers and prevents their polymerization.

Tropo-myosin

Tropo-myosin is a double-stranded, α-helical protein, with a molecular weight of about 70kD. It is intervened between the two helical strands of F actin, i.e. tropo-myosin molecules attach end to end and form a thin filament on each strand of the actin molecule.

One tropo-myosin molecule extends over seven actin molecules whereas 300-400 actin molecules are contained in each filament of the skeletal muscle (which is about 1 μm in length).

Troponin

Troponin is a spherical molecule, with a molecular weight of about 70 kD. It is positioned at regular intervals along the actin filaments.

Troponin is composed of three different types of subunits, called as troponin-T, troponin-I and troponin-C.

Troponin-T : It binds the other two subunits to the tropomyosin molecules, non-covalently.

Tropmyosin-troponin complex in turn regulates muscle contraction, by controlling access of the myosin heads on the thick filaments, to their binding sites on the actin.

Troponin-I : It is the inhibitory subunit. Its function is to inhibit the interaction of the head of myosin with actin, thereby preventing contraction of myofibrils.

Troponin-C : It is a Ca^{2+} binding protein and is rich in aspartate and glutamate.

Troponin C can also competitively bind Mg^{2+} ions.

Thin filaments consist of chains of the actin molecules with the associated tropomyosin and troponin molecules.

Q. Describe various biochemical changes that occur during muscle contraction.

Ans. Impulses, transmitted via motor nerve (across the neuromuscular junction to the sarcolemma of the muscle fibre), initiates contraction of myofibrils. This in turn stimulates the release of Ca^{2+} ions which bind to troponin-C of the thin actin filament and produces a configurational change in it. This is followed by shift of the tropo-myosin and leads to attachment of cross-bridges of the thick myosin filaments to the thin actin filaments.

ATP binds to special receptor sites on the surface of the myosin. The tendency for ATP to bind, is so efficient that under normal conditions, almost every myosin head is bound to ATP. Formation of the charged myosin.ATP-complex alters conformation of the protein in a manner that the charged complex has a greater tendency to bind to an actin molecule. After myosin.

ATP-complex binds to actin, ATP binds to a myosin head whose ATPase action causes hydrolysis of ATP. This in turn puts the myosin head into a high-energy conformation. The myosin head is now weak to bind, to actin monomer that is closer to the A line. This in turn results in the release of inorganic phosphate (Pi). The release of Pi further causes a conformational shift and increases the affinity of myosin for actin. The resulting transient state is immediately followed by the power stroke; causing a further conformational shift that sweeps myosin heads C-terminal tail towards the Z line, ADP is then released, thereby completing the cycle.

Myosin heads walk or row-up the adjacent thin filaments, towards the Z line with the concomitant, contraction of the muscle, i.e. the actin filaments slide along the thick myosin filaments (which otherwise were stationary), towards the centre of the sarcomere, thereby shortening its size. These events occur so rapidly that the whole process of muscle contraction takes only 10-20 milliseconds.

At the end of the neuro-musculatory phase, Ca^{2+} ions are drawn back into the sacroplasmic reticulum network and cross-bridges of the myosin retract. As a result of it, the sarcomeres and the muscle fibres relax and return to the resting state.

In the low-energy state, the complex remains intact until a new molecule of ATP is bound, usually occurring within a very short time (about 1 millisecond in a normal muscle).

The sequence continues, depending upon the supply of ATP and other components, essential for the muscle contraction.

Control of Muscle Contraction

The stimulation of a myofibril by nerve impulse, as mentioned above, results in an influx of Ca^{2+} from the sarcoplasmic reticulum. As a result of it, intracellular Ca^{2+} concentration increases. At higher concentration, additional Ca^{2+} binds to troponin C, thereby causing a conformational change in the troponin-tropomyosin complex. The movement of Ca^{2+} exposes the site on actin where myosin head binds. When myofibril Ca^{2+} concentration is low, Ca^{2+} is specifically ejected from the myofibril, by ATP-driven protein pump. The troponin-tropomyosin complex then assumes its resting conformation which in turn blocks myosin binding to actin and thereby causing the muscle to relax.

Q. How ATP is regenerated in the muscle ?

Ans. The energy for muscle contraction is provided by ATP, which is hydrolysed to ADP and Pi.

ATP can be regenerated, in the muscle by several routes:

1. ***By Lohman's reaction:*** ATP can be regenerated by utilizing high-energy phosphate compound, creatine phosphate. This reaction is catalyzed by creatine kinase (creatine phosphokinase or CPK) and is called as Lohman's reaction.

$$\text{ADP + Creatine-phosphate} \xleftrightarrow{\text{Creatine kinase}} \text{ATP + Creatine}$$

2. ***By adenylate kinase reaction:*** ATP can also be regenerated from ADP by the enzyme, adenylate kinase which is also referred to as myokinase.

$$\text{ADP + ADP} \xleftrightarrow{\text{Adenylate kinase}} \text{ATP + AMP}$$

3. ***From anaerobic glycolysis :*** Muscles also possess the capacity to synthesize ATP under anaerobic conditions by utilizing its glycogen stores.

A limited supply of oxygen, i.e. a short period of strenuous exercise, in turn results in the accumulation of large amount of lactate, in the blood.

4. ***By oxidative processes :*** Prolonged exercise gradually increases oxygenation of the muscle to a maximum. Thus ATP may also be produced by various oxidative processes, such as from aerobic glycolysis followed by the citric acid cycle, and from β-oxidation of fatty acids.

Q. Describe various types of muscle fibres.

Ans. Human muscles generally contain two types of the muscle fibres. These muscle fibres contain different isomers of the myosin molecules whose relative proportions, however, vary with the exercise, electric stimulation and growth:

1. ***Type I muscle fibres :*** These are also referred to as the red muscle fibres. They have a large number of mitochondria and thus have high oxidative capacity. They can regenerate ATP by aerobic metabolism of, both glucose and fatty acids. These muscle fibres are also resistant to fatigue.

2. ***Type II muscle fibres :*** These are also referred to as white muscle fibres.

These muscle fibres possess very few mitochondria and thus have a low oxidative capacity. These can regenerate ATP by metabolizing glycogen stores, by the process of anaerobic glycolsis.

Q. What is muscular dystrophy ?

Ans. Muscular dystrophy refers to a group of **hereditary progressive diseases,** each with unique phenotypic and genetic features, which cripple and kill children, such as **Duchenne muscular dystrophy.**

This is an X-linked recessive disorder present at birth but usually becomes apparent between ages 3 and 5. By age 5, muscle weakness is obvious by muscle testing. The disease is **caused by a mutation of the gene that encodes a protein dystrophin** and is localized to the inner surface of sarcolemma of the muscle fiber.

Serum creatine kinase (CPK) **levels are invariably elevated** to between 20 and 100 times normal. The levels are abnormal at birth but decline late in the disease because of inactivity and loss of muscle mass.

21

Hemoglobin Metabolism

Q. What is hemoglobin ? Describe its functions.

Ans. Hemoglobin is a globular protein, which is present in high concentration in red blood cells. A molecule of hemoglobin consists of four heme groups along with the four-polypeptide chains, which are synthesized separately and subsequently, bind to four heme groups. It binds oxygen in the lungs and transports it to different cells in the body.

FUNCTIONS OF HEMOGLOBIN

- Most important function of hemoglobin is the transport of gases, i.e. O_2 from the lungs to the tissues, and CO_2 from the tissues to the lungs. Binding of O_2 with hemoglobin is referred to as oxygenation while delivery of O_2 from oxyhemoglobin is designated as deoxygenation.
- Hemoglobin also **transports CO_2 and H^+** from the tissues to the lungs.
- It also **carries and releases nitric oxide (NO) in the blood vessels of the tissues.**

Q. What is Bohr effect ?

Ans. Binding of O_2 to the Fe^{2+} triggers conformational change in the hemoglobin molecule. This is also referred to as the relaxed (R) conformational state.

After delivery of O_2 sixth coordinate position of the Fe^{2+} atom becomes vacant (unoccupied). This is called deoxyhemoglobin. Deoxy- conformation of the hemoglobin is referred to as tense (T) conformational state.

T and R conformational equilibrium of hemoglobin controls delivery of O_2 to the tissues and transport of CO_2 away from the tissues. This equilibrium between the T and the R state is regulated by H^+ concentration. H^+ is released when T conformation is converted to R.

Thus, affinity of oxygen for hemoglobin is affected by pH. It decreases as the pH becomes more acidic.

Increase in acidity of hemoglobin (as it binds O_2), or increase in basicity (as it releases oxygen) is referred to as Bohr effect. It may be expressed by the following equation:

$$H.Hb + O_2 \rightleftharpoons HbO_2 + H^+$$

It suggests that an increase in H^+ concentration favors release of oxygen from oxyhemoglobin. Conversely, oxygenation of hemoglobin lowers pH of the solution.

Q. What do you mean by isohydric transport of CO_2 ?

Ans. Metabolizing cells utilize O_2 and produce CO_2, which diffuses from the cells of the tissues to red blood cells. Within red blood cells, CO_2 is dissolved in water and converted to H_2CO_3, by the enzyme carbonic anhydrase. Spontaneously, carbonic acid is dissociated to HCO_3^- and H^+. H^+, which dissociates from carbonic acid, binds to deoxyhemoglobin, forcing the equilibrium from oxyhemoglobin to deoxyhemoglobin with the dissociation of O_2, which diffuses out of the red blood cells of the tissues. HCO_3^- is transported (in plasma) to the lungs.

In the lungs, high O_2 pressure forces the reaction in the opposite direction. It promotes conversion of deoxyhemoglobin to oxyhemoglobin. During this process, H^+ are dissociated and combine with HCO_3^- to reform carbonic acid. Carbonic anhydrase now dissociates H_2CO_3 to CO_2 and H_2O. CO_2 diffuses out of the blood, into the lung alveoli, and is expired into air.

About 80% of the cellular CO_2 is transported from the tissues to the lungs by this mechanism, i.e. as HCO_3^-. This is called isohydric transport of CO_2.

Q. What is carbaminohemoglobin ?

Ans. From the cells of the tissues CO_2 diffuses into the red blood cells, in capillaries of tissues, and reacts with the NH_2-terminal amino groups of hemoglobin polypeptide chains forming carbaminohemoglobin.

This reaction also produces H^+, which promotes dissociation of O_2 from hemoglobin that is diffused out of the red blood cells to the tissues.

$$\text{Hb-N}\begin{matrix}\text{H}\\ \\ \text{H}\end{matrix} + CO_2 \rightleftharpoons \text{Hb-N}(\text{H})\text{-C}(=\text{O})\text{-O}^- + H^+$$

In the lungs, high O_2 pressure forces these reactions in the opposite direction. This, in turn, leads to the expiration of CO_2 from lungs. Free H^+ thus, promotes dissociation of the carbamino-group from Hb, to reform CO_2, which is expired from lungs.

Q. Describe the role of hemoglobin in the transport of nitric oxide.

Ans. Nitric oxide (NO) is a potent vasodilator. It has a very short life in the blood but is prevented from rapid destruction by hemoglobin.

Hemoglobin reversibly binds NO. As hemoglobin changes from the oxy (R) conformation to the deoxy- (T) conformation, it dissociates the bound NO to glutathione (GSH). NO is thus, released as glutathione-nitric oxide complex (GS-NO). Glutathione not only stabilizes NO against its rapid degradation but also delivers it efficiently to its receptors in the cells of the blood vessel wall and promotes relaxation of the vascular wall. This, in turn, facilitates the transfer of gases between blood and cells of the tissues.

Q. Describe the process of heme synthesis. How it is regulated?

Ans. Heme consists of one atom of ferrous iron (Fe^{2+}) and a tetrapyrrole ring, called protoporphyrin IX.

The organic portion of the heme is derived from eight residues of glycine and succinyl CoA. In the biosynthesis of heme, some of the reactions occur in mitochondria while others in the cytosol.

1. In the first step, succinyl CoA condenses with glycine and forms δ-aminolevulinic acid (ALA), in the mitochondria.

$$\text{Glycine + Succinyl CoA} \xrightarrow[\text{PLP}]{\text{ALA synthase}} \delta\text{-Aminolevulinic acid} + \text{CoA} + CO_2$$

This reaction is catalyzed by the enzyme δ-aminolevulinic acid synthase (ALA synthase), which requires pyridoxal phosphate (PLP) as a coenzyme. This is the rate-limiting step of heme synthesis.

Increased levels of ALA synthase along with the reduced levels of some enzymes of the heme biosynthetic pathway result in a group of disorders referred to as **porphyrias.**

2. In the next step, two molecules of δ-aminolevulinic acid condenses, asymmetrically, and form porphobilinogen.

This reaction is catalyzed by the enzyme aminolevulinic acid dehydratase (ALA dehydratase).

$$\delta\text{-Aminolevulinic acid} \xrightarrow[\searrow 2H_2O]{\text{ALA-dehydratase}} \text{Porphobilinogen}$$

ALA dehydratase is a zinc-containing enzyme, which occurs in the cytosol. It consists of eight subunits, four of which interact with the substrate.

ALA dehydratase is sensitive to heavy metals particularly, lead. ALA levels increase in lead poisoning.

3. Porphobilinogen undergoes deamination, by the enzyme porphobilinogen deaminase (hydroxymethylbilane synthase or uroporphyrinogen I synthase), and is converted to a linear tetrapyrrole, called hydroxymethylbilane.

$$\text{Porphobilinogen} \xrightarrow{\text{Porphobilinogen deaminase}} \text{Hydroxymethylbilane}$$

Decreased activity of porphobilinogen deaminase along with the increased activity of ALA synthase results in a condition called acute intermittent porphyria.

4. Spontaneously, with ring closure, hydroxymethylbilane is converted to uroporphyrinogen I.

Alternatively, uroporphyrinogen III synthase converts hydroxymethylbilane to uroporphyrinogen III.

Only, type I and type III isomers are found in nature.

Porphyrins having symmetric arrangement of the substituent groups on four-pyrrol rings are designated as type I while those having asymmetric substitutions are referred to as type III.

Under normal conditions, uroporphyrinogen formed is, almost exclusively, type III. Heme is also a type III porphyrin.

Abnormality of the reticulocyte uroporphyrinogen III synthase causes an inherited disease called **erythropoietic porphyria.** It is associated with marked cutaneous light sensitization, and increased synthesis of uroporphyrinogen I and coproporphyrinogen I in the bone marrow.

5. Uroporphyrinogen I and III are converted to the respective coproporphyrinogen by the enzyme uroporphyrinogen decarboxylase.

Iron salts inhibit uroporphyrinogen decarboxylase.

Genetic deficiency of uroporphyrinogen decarboxylase leads to a disease, known as porphyria cutanea tarda, which shows cutaneous manifestations, primarily, with sensitivity to light. The manifestation is expressed in-patients who take drugs that cause an increase in porphyrin synthesis, or in individuals who consume large amount of alcohol. This, in turn, leads to accumulation of iron, which further inhibits uroporphyrinogen decarboxylase.

6. In the next reaction, coproporphyrinogen III enters the mitochondria and is converted to protoporphyrinogen IX.

This reaction is catalyzed by the enzyme coproporphyrinogen oxidase, which requires molecular oxygen. This enzyme does not act on coproporphyrinogen I.

(These porphyrins are identified as IX since they were designated ninth in a series of the postulated isomers.)

Inherited deficiency of the enzyme coproporphyrinogen oxidase leads to a form of the hepatic porphyria, known as **hereditary coproporphyria.**

7. Protoporphyrinogen oxidase acts on protoporphyrinogen IX and converts it to protoporphyrin IX, in the mitochondria.

$$\text{Protoporphyrinogen IX} \xrightarrow{\text{Protoporphyrinogen oxidase}} \text{Protoporphyrin IX}$$

Protoporphyrin IX is water insoluble. Its excessive amounts are excreted by the biliary system into the intestinal tract. Deficiency of the enzyme, protoporphyrinogen oxidase, leads to a condition called **variegate porphyria.**

8. In the final step of heme synthesis, ferrous iron (Fe^{2+}) is incorporated into protoporphyrin IX. This reaction is catalyzed by the enzyme ferrochelatase, which requires some reducing substances.

$$\text{Protoporphyrin IX} \xrightarrow[Fe^{2+}]{\text{Ferrochelatase}} \text{Heme}$$

This enzyme is sensitive to heavy metals especially, lead. It is also sensitive to iron deprivation.

Inherited deficiency of the enzyme ferrochelatase results in a disease referred to as hereditary **protoporphyria.**

Regulation of Heme Biosynthesis

ALA synthase occurs in two forms. These are called hepatic (ALA-S1) and erythrocytic (ALA-S2) enzymes.

ALA-S1 controls the rate-limiting step of heme synthesis. Further, heme acts both, as a repressor of synthesis of the enzyme as well as its inhibitor.

Q. What is porphyria? Describe various types of porphyrias.

Ans. Porphyria is a disorder due to abnormality in the pathway of heme biosynthesis. It may be genetic or acquired. Six major types of porphyria have been described, five of which are inherited as autosomal dominant:

Derangement in Porphyrin Metabolism in Different Types of Porphyria

Porphyria	Enzyme defect	Major symptoms	Laboratory findings
Acute intermittent porphyria	Increased ALA synthase, decreased porpho-bilinogen deaminase	Abdominal pain, neuro-psychiatric symptoms	Urinary porpho-bilinogen (+)
Erythropoietic porphyria	Decreased uropor-phyrinogen III synthase	No photo-sensitivity	Porpho-bilinogen (-), uropor phyrin (+)

Porphyria cutanea tarda	Decreased uroporphyrinogen decarboxylase	Photo sensitivity	Porphobilinogen (-), uroporphyrin (+)
Hereditary coproporphyria	Increased ALA synthase, decreased coprophyrinogen oxidase	Photo sensitivity, abdominal pain, neuropsychiatric symptoms	Urinary porpho bilinogen (+), urinary uroporphyrin (+), fecal protoporphyrin (+)
Variegate porphyria	Increased ALA synthase, decreased coproporphyrinogen oxidase	Photo-sensitivity, abdominal pain, neuro-psychiatric symptoms	Urinary porphobilinogen (+), fecal protoporphyrin (-)
Hereditary protoporphyria	Decreased ferrochelatase	Photo-sensitivity	Fecal protoporphyrin (-), red cell protoporphyrin (+)

Q. Describe the process of heme degradation.

Ans. Red blood cells have limited life span of approximately 100-120 days. These senescent cells are recognized by their membrane change, removed, and engulfed by the reticuloendothelial system, at the extra-vascular site. Degradation of red blood cell occurs in spleen, bone marrow, liver, and lymph glands.

Degradation of heme occurs mainly, in the liver. If degradation of red blood cells occurs in the tissues other than the liver, hemoglobin is transported to the liver by means of haptoglobulins.

After the aged red blood cells are recognized by macrophages, they are rapidly engulfed by the phagocytes and form phagosomes. They fuse with the primary lysosomes and form secondary lysosomes. Lysosomal cathepsin results in complete degradation of the cellular proteins, including globin of hemoglobin, to the constituent amino acids, which are utilized for general metabolic needs.

Heme is degraded in the reticuloendothelial cells, to a linear tetrapyrrole (biliverdin IXα), by the microsomal enzyme system, which is designated as heme oxygenase. This enzyme requires molecular oxygen and NADPH, and is induced by heme. Heme oxygenase catalyzes the cleavage of α-methenyl bridge, which is quantitatively converted to carbon monoxide (CO) that is trapped by the hemoglobin and eventually exhaled.

Biliverdin is reduced to bilirubin by the enzyme biliverdin reductase.

One gram of hemoglobin yields about 35 mg of bilirubin.

Iron, which is released from the porphyrin ring, is transferred from the cells by an iron transport protein, called transferrin, and is made available for resynthesis of hemoglobin. Besides the senescent red blood cells, bilirubin is also derived from the turnover of the other heme containing proteins, such as cytochromes.

Q. Describe various types of normal and abnormal hemoglobins.

Ans.

NORMAL HEMOGLOBINS

Hemoglobin normally found in the body that performs all the physiologic functions is referred to as the normal hemoglobin. Different types of human hemoglobins are:

Adult Hemoglobin (HbA_1) : This is the major hemoglobin present in adult human beings and consists of two α chains and two β chains ($\alpha_2\beta_2$).

Minor Adult Hemoglobin (HbA_2) : Minor adult hemoglobin comprises about 2% of total hemoglobin in adults. It consists of two α chains and two delta (δ) chains ($\alpha_2\delta_2$) instead of two α and two β chains present in HbA.

Fetal Hemoglobin (HbF) : It predominates in the blood of human fetus and early embryo HbF contains two α chain (as in HbA_1) along with two γ chains ($\alpha_2\gamma_2$). γ Chain differs in amino acid sequence from the β chain of HbA_1.

Two other forms of the fetal hemoglobin appear in the embryo (in the first month after conception). These hemoglobins contain two zeta (ξ) chains with either two epsilon (ε) chains ($\xi_2\varepsilon_2$) or two γ chains ($\xi_2\gamma_2$). These are called Hb.Gower-1 or Hb.Portland, respectively.

ABNORMAL HEMOGLOBINS

There are several mutants of hemoglobin, called **abnormal hemoglobin.**

A mutation may cause a change in hemoglobin structure, increase or decrease its O_2 affinity, or increase rate of oxidation of the heme iron from Fe^{2+} to Fe^{3+}.

Abnormal hemoglobin may arise by substitution of one amino acid by another, a peptide chain by another, which may also be either too large or too small, or by the placement of a charged (polar) group inside of a domain. These abnormal hemoglobins get denaturated and precipitate within the erythrocytes, forming Heinz bodies, which can even damage the erythrocyte membrane.

Some of the Abnormal Hemoglobins are:

Methemoglobin (HbM): It is non-functional hemoglobin, which has Fe^{3+} instead of Fe^{2+}. Patients with high concentration of methemoglobin show cyanosis (bluish color of the skin).

Hemoglobin S (HbS) : It is a variant of HbA_1 in which substitution occurs in the sixth position of the β-globin chain of HbA_1. As a result of it, HbS gets polymerized and precipitate within the red blood cells. Red blood cells assume sickle shape and result in anemia, designated as sickle-cell anemia.

Hemoglobin C: **Hemoglobin C** is another variant of HbA_1 where glutamic acid, normally present at the sixty position of the β-chain of HbA_1 gets replaced by lysine. This, in turn, results, in the formation of aggregates and reduces survival time for red blood cells. This form of hemoglobinopathy exhibits more limited pathological effects and is commonly found among certain black African population.

Q. What is sickle cell anemia ?

Ans. This is a consequence of point mutation. Red blood cells of such individuals have hemoglobin S (HbS) which differs from that of the normal adult hemoglobin (HbA_1) with respect to only one amino acid in their β-chains, i.e. glutamic acid normally present at the sixth position of the β chain of HbA_1 gets replaced with valine in HbS. This is due to a change in a single nucleotide within the codon (from GAA or GAG, which code for glutamic acid to GUA or GUG, which code for valine). This missense mutation results in a mutant gene, which is present in homozygous state and results in sickle cell disease. Such a mutation is mostly seen in some blacks, (in some parts of the world), particularly where malaria is widespread. Hemoglobin S though can bind and release oxygen but abnormally. Heterozygous individuals do not exhibit symptoms of sickle cell anemia except under extreme conditions of hypoxia, due to the reason that such individuals have 50% of HbA_1 and 50% of HbS in their red blood cells. HbS can be detected by gel electrophoresis. Sickle cell anemia can also be diagnosed by recombinant DNA techniques.

Q. What is hemoglobinopathy ?

Ans. Patients with unstable hemoglobin develop a condition called hemoglobinopathy with characteristic features, such as anemia, reticulocytosis, splenomegaly and urobilinuria. Hemoglobinopathies are a heterogeneous group of single-gene disorders.

A mutant gene may result in the substitution of one amino acid in one type of the polypeptide chain by another amino acid, or loss of one or more amino acid residues of a polypeptide chain. This, in turn, results in the formation of structurally abnormal hemoglobin. In some cases, the change may be clinically insignificant while in others it may cause a serious disease. The most common hemoglobinopathy is sickle cell disease.

Q. Write a note on thalassemia.

Ans. Thalassemia is a family of related genetic disorders that arise due to the deletion of one or more globin-like genes in either the globin gene cluster or a defect in transcription, and/or processing of mRNA of the globin gene. If there is reduced synthesis or total lack of synthesis of α-globin mRNA, the disease is classified as α-thalassemia. On the other hand, if β-globin mRNA level is affected, it is called β-thalassemia.

One to four α-globin genes may be missing in the patients

with α-thalassemia. If one α-globin gene is missing, the condition is called as α-thalassemia 1 (α-thal 1). When two α-globin genes are missing, it is called α- thal 2. Both the conditions are associated with mild to moderate anemia. On the other hand, if three α-globin genes are missing it results in the synthesis of more β-globin molecules, forming a tetramer containing four β - globin sub-units. This condition is called as HbH disease. When all the four α-globin genes are absent, it results in a fatal condition called hydrops fetalis.

β-thalassemias also exhibits different degree of severity and can be caused by a variety of defects or deletions.

Index

A

B

C

D

E

S

T

U